Contents

	Page
Preface	v
Introduction	vi
Scientific—Common Names of Nematodes and Platyhelminths	1
Scientific—Common Names of Annelids	6
Scientific—Common Names of Arthropods	7
Scientific—Common Names of Molluscs	71
Common—Scientific Names of Nematodes and Platyhelminths	73
Common—Scientific Names of Annelids	78
Common—Scientific Names of Arthropods	80
Common—Scientific Names of Molluscs	146

Ministry of Agriculture, Fisheries and Food

Invertebrates of Economic Importance in Britain

Common and Scientific Names

*formerly Technical Bulletin No. 6
Common Names of British Insect and other Pests*

Compiled by Paul R. Seymour
Harpenden Laboratory, Harpenden

London: Her Majesty's Stationery Office

© *Crown copyright 1989*
First published 1957 (as Technical Bulletin 6
Common Names of British Insect and Other Pests)

Fourth Edition 1989

ISBN 0 11 242829 0

HMSO publications are available from:

HMSO Publications Centre
(Mail and telephone orders only)
PO Box 276, London, SW8 5DT
Telephone orders 01-873 9090
General enquiries 01-873 0011
(queuing system in operation for both numbers)

HMSO Bookshops
49 High Holborn, London, WC1V 6HB 01-873 0011 (Counter service only)
258 Broad Street, Birmingham, B1 2HE 021-643 3740
Southey House, 33 Wine Street, Bristol, BS1 2BQ (0272) 264306
9–21 Princess Street, Manchester, M60 8AS 061-834 7201
80 Chichester Street, Belfast, BT1 4JY (0232) 238451
71 Lothian Road, Edinburgh, EH3 9AZ 031-228 4181

HMSO's Accredited Agents
(see Yellow Pages)

and through good booksellers

Preface

This work was first published under the title 'Common Names of British Insect and Other Pests' in two parts by the Association of Applied Biologists (Part 1 in 1947; Part 2 in 1952). Entries were restricted to British pests of plants, domestic animals, stored products and timber. The original lists were compiled by a joint sub-committee of the MAFF Conference of Advisory Entomologists and the Pests and Diseases Committee of the Association of Applied Biologists.

A revision by I. Thomas, H. W. Janson and Audrey A. Aitken was published in 1957 by Her Majesty's Stationery Office as MAFF Technical Bulletin No. 6. A further revision, largely prepared by H. W. Janson, was produced in 1968 under the same authorship.

In 1979 the work was updated by the present compiler; platyhelminths and annelids were included for the first time and the arthropod and nematode lists were extended.

The main purpose of this latest edition has been to bring the scientific names up to date. At the same time the opportunity has been taken to include a number of native species now regarded as pests and some foreign pests which, since 1979, have been more frequently intercepted in Britain.

In preparing the present edition I wish to thank the ADAS Regional Entomologists and the many other people who have given valuable advice and have recommended names for inclusion, especially Mr Richard Adams of the Storage Pests Department, Slough Laboratory, and Mr A. C. Kirkwood of the Central Veterinary Laboratory, Weybridge.

I am also grateful to the many people who have helped by checking the validity of the scientific names, in particular Dr Laurence Mound, Keeper of the Department of Entomology, British Museum (Natural History), and many members of his staff; Mr Donald Macfarlane, Commonwealth Institute of Entomology; Dr R. L. S. Muller, Commonwealth Institute of Parasitology; the late Dr Alan R. Stone, Nematology Department, Rothamsted Experimental Station; and Dr N. J. Evans and Dr R. W. Sims, both of the Department of Zoology, British Museum (Natural History).

I would also like to thank Miss Mary M. Davis and Mr Simon T. R. Jeavons of Harpenden Laboratory for diligence in checking copies of the typescript. Finally I would like to express my thanks to Miss Marion Gratwick and Dr David V. Alford, both of ADAS Entomology Discipline, for helpful suggestions and encouragement at all stages.

Paul R. Seymour

Introduction

The layout of this book follows that of the third edition (1979).

The following categories are included:

(1) pests and beneficial species important in agriculture, horticulture and forestry,
(2) pests of stored products and timber, and
(3) the more common pests of veterinary, medical and public health importance.

A number of species are regarded as both pest and beneficial, e.g. wasps. A few others are listed which are neither harmful nor beneficial but are often thought to be injurious or troublesome. Entries are restricted to metazoan invertebrates that occur in Britain and include foreign species occasionally encountered, e.g. in association with imported produce.

Scientific Names

The surname of the author of the first published description of the species is given immediately after the scientific name. Author's surnames are given in full except for Linnaeus whose name is abbreviated to L. The author's name appears in brackets when the generic name differs from the one used in the original description. The reviser, i.e. the person responsible for the generic recombination, is included only for the plant-parasitic nematodes, in accordance with nematological practice, e.g.

Radopholus similis (Cobb) Thorne,

indicating that Cobb was the author and Thorne the reviser.

In the list of Scientific—Common Names, rejected scientific names, i.e. rejected synonyms or homonyms (more usually referred to as 'synonyms'), are inset; the appropriate cross reference is given, e.g.

Typhlocyba douglasi Edwards, see *Fagocyba cruenta*

Misidentifications, i.e. names wrongly applied to species, and misspellings are also included, inset. Rejected scientific names, misidentifications and misspellings are restricted to those occurring at least fairly frequently in literature of the past 50 years.

In the list of Common—Scientific Names, rejected scientific names, misidentifications and misspellings are placed beneath the scientific name, inset and prefixed 'syn.', 'misident.' and 'misspelling' respectively, e.g.

willow flea beetle . . . . *Chalcoides aurata* (Marsham)
    syn. *Crepidodera aurata* (Marsham)
    misident. *C. helxines* (L.)

wingless weevils . . . . *Otiorhynchus* spp.
    misspelling *Otiorrhynchus*

Common Names

Some names starting with adjectives that occur frequently in the list, e.g. common, lesser, yellow, are listed twice. The main entry places the first word (or occasionally two words) last, while the secondary entry (inset and cross-referenced) places the words in their correct sequence, e.g.

gooseberry sawfly, common . . . . *Nematus ribersii* (Scopoli)
  common gooseberry sawfly, *see* gooseberry sawfly, common

Names that do not indicate the order or group to which the species belong are followed by an appropriate collective term in brackets, e.g.
    sheep ked (fly)

In the list of **Scientific—Common Names**, when there are two or more generally accepted common names they are listed alphabetically immediately below one another, e.g.

Contarinia merceri Barnes . . . . . . { cocksfoot midge
                                                  { foxtail midge

Alternative but less favoured common names that occur in literature of the past 50 years are listed, inset, below the accepted name, e.g.

                        corn moth
                          European grain moth
                        wolf moth

Common names applying to an immature stage are prefixed by an appropriate juvenile term, e.g.

Phytomyza ilicis Curtis . . . . . . *larva* = holly leaf miner

When there are different common names for the adult and immature stages the juvenile name is placed below that of the adult, e.g.

Athous haemorrhoidalis (Fabricius) . . garden click beetle
                                      larva = garden wireworm

In the list of **Common—Scientific Names**, less favoured common names are inset and cross-referenced to the generally accepted name(s), e.g.

    debris bug, *see* stack bug
  grocers' itch mite, *see* cosmopolitan food mite
                and house mite

When the common name applies to an immature stage this stage is indicated, e.g.
    tanbark borer . . . . larva of *Phymatodes testaceus* (L.)

When there are different common names for the adult and immature stages the juvenile name is listed twice—by itself and inset below the adult name, e.g.

    pine looper . .   . .   larva of *Bupalus piniaria* (L.)

    bordered white moth   . .   . .   . .   *Bupalus piniaria* (L.)
       larva = pine looper

When, however, the adult and immature names follow one another in the list or are separated by no more than a few entries, the juvenile name is directly referred to the adult entry, e.g.

    apple fruit miner, *larva* of apple fruit moth

Symbols

The meaning of symbols prefixing entries is as follows:

* foreign species which are found from time to time (e.g. associated with imported produce) and occasionally breed in this country.

‡ species extinct in Britain.

† beneficial species.

Scientific — Common Names of Nematodes and Platyhelminths

Scientific Name	Common Name
Acanthocephala spp.	thorny-headed worms
see also *Polymorphus* spp.	
Ancylostoma spp.	hookworms
Anguina spp.	flower and leaf-gall nematodes
see also *Subanguina* spp.	
Anguina agrostis (Steinbuch) Filipjev	bent-grass seed nematode
Anguina graminis (Hardy) Filipjev	fescue leaf-gall nematode
Anguina graminophila (T. Goodey) Christie, see *Subanguina graminophila*	
Anguina millefolii (Löw) Filipjev	yarrow leaf-gall nematode
Anguina radicicola (Greeff) Teploukhova, see *Subanguina radicola*	
‡*Anguina tritici* (Steinbuch) Chitwood	wheat gall nematode
not found since c. 1956	ear cockles eelworm
Aphelenchoides blastophthorus Franklin	scabius bud nematode
Aphelenchoides composticola Franklin	mushroom spawn nematode
see also *Ditylenchus myceliophagus*	
Aphelenchoides fragariae (Ritzema Bos) Christie	leaf nematode
see also *Aphelenchoides ritzemabosi*	bud and leaf nematode
	fern eelworm
Aphelenchoides olesistus (Ritzema Bos) Steiner, see *Aphelenchoides fragariae*	
Aphelenchoides ribes (Taylor) T. Goodey, see *Aphelenchoides ritzemabosi*	
Aphelenchoides ritzemabosi (Schwartz) Steiner & Buhrer	chrysanthemum nematode
	blackcurrant eelworm
	bud and leaf nematode ⎫ (see also
	leaf nematode ⎭ *Aphelenchoides fragariae*)
Aphelenchoides subtenuis (Cobb) Steiner & Buhrer	narcissus bulb and leaf nematode
Ascaridia galli (Shrank)	poultry roundworm
see also *Heterakis gallinarum*	
Ascaris lumbricoides L.	human large roundworm
Ascaris megalocephala (Goeze), see *Parascaris equorum*	
Ascaris suum Goeze	pig roundworm
Bidera avenae (Wollenweber) Krall & Krall, see *Heterodera avenae*	
Bothriocephalus latus (L.), see *Diphyllobothrium latum*	
Bursaphelenchus xylophilus (Steiner & Buhrer) Nickle	pine wood nematode
Cactodera cacti (Filipjev & Schuurmans Stekhoven) Krall & Krall	cactus cyst nematode
	cactus root eelworm

SCIENTIFIC NAMES—Nematodes and Platyhelminths

Scientific Name	Common Name
Capillaria spp.	poultry nematodes
CESTODA	tapeworms
Cooperia spp. see also *Nematodirus* spp. & *Trichostrongylus* spp.	intestinal worms
CRICONEMATINAE	ring nematodes

Cysticercus cellulosae (Gmelin), see *Taenia solium*

Davainea proglottina (Davaine) Blanchard	poultry tapeworm

Dicrocoelium dendriticum Rudolphi, see *Dicrocoelium lanceolatum*

Dicrocoelium lanceolatum	lancet fluke
Dictyocaulus spp.	lungworms
Diphyllobothrium latum (L.)	broad fish tapeworm (of man) broad tapeworm fish tapeworm
Dirofilaria immitis (Leidy)	dog heartworm heartworm (of dog)
Ditylenchus destructor Thorne	potato tuber nematode iris bulb nematode tuber-rot eelworm
Ditylenchus dipsaci (Kühn) Filipjev	stem nematode bulb eelworm stem and bulb eelworm
Ditylenchus myceliophagus J. B. Goodey see also *Aphelenchoides composticola*	mushroom spawn nematode

Ditylenchus radicicolus (Greeff) Filipjev, see *Subanguina radicicola*

Echinococcus spp.	dwarf dog tapeworms *larvae* = hydatid cysts
Enterobius vermicularis (L.)	pinworm (of man) seatworm
Fasciola hepatica (L.)	liver fluke
Globodera spp.	round-cyst nematodes
Globodera achilleae (Golden & Klindić) Behrens	yarrow cyst nematode milfoil cyst nematode
Globodera pallida (Stone) Behrens	white potato cyst nematode pale potato cyst nematode potato root eelworm (see also entry below)
Globodera rostochiensis (Wollenweber) Behrens	yellow potato cyst nematode golden nematode potato root eelworm (see also entry above)
Haemonchus contortus (Rudolphi) Cobb	stomach worm barber's pole worm
Helicotylenchus spp. see also *Rotylenchus* spp.	spiral nematodes

SCIENTIFIC NAMES—Nematodes and Platyhelminths

Scientific Name	Common Name
Hemicycliophora spp.	sheath nematodes
Heterakis gallinae Gmelin, see *Heterakis gallinarum*	
Heterakis gallinarum (Schrank)	poultry roundworm
see also *Ascaridia galli*	
HETERODERINAE	cyst nematodes
	cyst eelworms
Heterodera spp.	lemon-shaped cyst nematodes
Heterodera achilleae Golden & Klindić, see *Globodera achilleae*	
Heterodera avenae Wollenweber	cereal cyst nematode
	cereal root eelworm
Heterodera cacti Filipjev & Schuurmans Stekhoven, see *Cactodera cacti*	
Heterodera carotae Jones	carrot cyst nematode
	carrot root eelworm
Heterodera cruciferae Franklin	brassica cyst nematode
	brassica root eelworm
	cabbage root eelworm
Heterodera fici Kir'yanova	fig cyst nematode
	fig root eelworm
Heterodera galeopsidis Goffart	hemp-nettle cyst nematode
	galeopsis cyst nematode
Heterodera goettingiana Liebscher	pea cyst nematode
	pea root eelworm
Heterodera humuli Filipjev	hop cyst nematode
	hop root eelworm
Heterodera maior O. Schmidt, see *Heterodera avenae*	
Heterodera major: misspelling, see entry above	
Heterodera pallida Stone, see *Globodera pallida*	
Heterodera punctata Thorne, see *Punctodera punctata*	
Heterodera rostochiensis Wollenweber, see *Globodera rostochiensis*	
note: the name *rostochiensis* was wrongly used in several published works prior to Stone, 1973, to describe *Globodera pallida*	
Heterodera schachtii Schmidt	beet cyst nematode
	beet eelworm
Heterodera trifolii Goffart	clover cyst nematode
	clover root eelworm
Heterodera urticae Cooper	nettle cyst nematode
Hoplolaimus spp.	lance nematodes
Hymenolepis nana (Siebold)	dwarf tapeworm
Hyostrongylus rubidus (Hassall & Stiles)	pig stomach worm
Longidorus spp.	needle nematodes
Meloidogyne spp.	root-knot nematodes
	root-knot eelworms
Meloidogyne arenaria (Neal) Chitwood	pea-nut root-knot nematode
Meloidogyne hapla Chitwood	northern root-knot nematode
Meloidogyne incognita (Kofoid & White) Chitwood	southern root-knot nematode
Meloidogyne javanica (Treub) Chitwood	Javanese root-knot nematode
Meloidogyne naasi Franklin	cereal root-knot nematode
MERLINIINAE	stunt nematodes
see also TYLENCHORHYNCHINAE	
Mermis nigrescens Dujardin	thunderworm
	rain worm

SCIENTIFIC NAMES—Nematodes and Platyhelminths

Scientific Name   Common Name

Metastrongylus spp. .. ..   pig lungworms
Moniezia spp. .. ..   sheep tapeworms
Multiceps spp. .. ..   larvae = coenurus cysts
Muellerius capillaris (Müller) .. ..   sheep lungworm

*Nacobbus spp. .. ..   false root-knot nematodes
NEMATODA ⎫
NEMATODEA ⎭ .. ..   nematodes
    eelworms ⎱ associated with plants
    roundworms (see also ⎫
      Toxascaris spp. and ⎬ associated with animals
      Toxocara spp.) ⎭
    threadworms
Nematodirus spp. .. ..   intestinal worms
  see also *Cooperia* and *Trichostrongylus*

Oesophagostomum dentatum (Rudolphi)   nodular worm
Oxyuris equi (Schrank) .. .. ..   horse pinworm
**Opisthorchis felineus* Rivolta .. ..   cat liver fluke
Ostertagia spp. .. ..   brown stomach worms

Paramphistomum spp. .. .. ..   rumen flukes
                          stomach flukes
Parascaris equorum Yorke & Maplestone   large roundworm (of horse)
Paratrichodorus spp. .. ..   stubby-root nematodes
  see also *Trichodorus* spp.
Paratylenchus spp. .. ..   pin nematodes
PLATYHELMINTHES .. .. ..   flatworms
Polymorphus spp. .. ..   thorny-headed worms
  see also *Acanthocephala* spp.
Pratylenchus spp. .. ..   root-lesion nematodes
                          meadow nematodes
Punctodera punctata (Thorne) Mulvey &
  Stone .. ..   grass cyst nematode

**Radopholus citropholus*
  Huettel, Dickson & Kaplan ⎫
**Radopholus similis* (Cobb) Thorne ⎭ ..   burrowing nematodes
Rotylenchus spp. .. ..   spiral nematodes
  see also *Helicotylenchus* spp.

**Stephanurus dentatus* Diesing .. ..   kidney worm (of pig)
                          swine kidney worm
Strongyloides papillosus (Wedl) .. ..   intestinal threadworm (of sheep)
Strongyloides ransomi Schwartz & Alicata   intestinal threadworm (of pig)
Subanguina spp. .. ..   flower and leaf-gall nematodes
  see also *Anguina* spp.
  Subanguina graminophila (T. Goodey)
    Brzeski .. ..   bent-grass leaf-gall nematode
Subanguina radicicola (Greeff) Paramonov   grass root-gall nematode
Syngamus spp. .. ..   gapeworms

SCIENTIFIC NAMES—Nematodes and Platyhelminths

Scientific Name	Common Name
Taenia spp.	*larvae* = bladder worms
Taenia multiceps Leske	*larva* = coenurus cyst
Taenia saginata Goeze	beef tapeworm
Taenia solium L.	pork tapeworm
Thelazia spp.	eyeworms
Toxascaris spp. } *Toxocara* spp. }	arrowhead worms roundworms (see also NEMATODA, NEMATODEA)
TREMATODA	flukes
Trichodorus spp. see also *Paratrichodorus* spp.	stubby-root nematodes
Trichostrongylus spp. see also *Cooperia* spp. and *Nematodirus* spp.	intestinal worms
Trichuris spp.	whipworms
Turbatrix aceti (Müller) Peters	vinegar nematode vinegar eel
TYLENCHORHYNCHINAE see also MERLINIINAE	stunt nematodes
Uncinaria stenocephala (Railliet)	dog hookworm
Xiphinema spp.	dagger nematodes

Scientific—Common Names of Annelids

Scientific Name	Common Name
†*Allolobophora chlorotica* (Savigny)	green worm
ANNELIDA	segmented worms
	true worms
†*Aporrectodea caliginosa* (Savigny)	grey worm
	nocturnal worm
†*Aporrectodea longa* (Ude)	long worm
Aporrectodea nocturna (Evans), see *Aporrectodea caliginosa*	
Dendrobaena subrubicunda (Eisen), see *Dendrodrilus rubidus subrubicunda*	
†*Dendrodrilus rubidus* (Savigny)	
ssp. *subrubicunda* (Eisen)	cockspur
	gilt tail
	yellow tail
†*Eisenia fetida* (Savigny)	brandling
	manure worm
Eisenia foetida (Savigny): misspelling, see entry above	
†*Eisenia rosea* (Savigny)	rosy worm
ENCHYTRAEIDAE	pot worms
	white worms
	aster worms
LUMBRICIDAE	earthworms
†*Lumbricus castaneus* (Savigny)	chestnut worm
	purple worm
†*Lumbricus rubellus* Hoffmeister	red worm
†*Lumbricus terrestris* L.	lob worm
	dew worm
	squirrel tail worm
	twachel

Scientific—Common Names of Arthropods

Scientific Name	Common Name
Abacarus hystrix (Nalepa)	cereal rust mite
Abax parallelepipedus (Piller & Mitterpacher)	parallel-sided ground beetle
Abraxas grossulariata (L.)	magpie moth currant moth
Acalitus brevitarsus (Fockeu)	alder erineum mite
Acalitus essigi (Hassan)	blackberry mite
Acalitus phloeocoptes (Nalepa)	plum spur mite plum leaf mite (see also *Aculus fockeui*)
Acalitus rudis (Canestrini)	witches' broom mite (of birch)

Acalla comariana (Zeller), see *Acleris comariana*

Acanthocinus aedilis (L.)	common timberman (beetle)

Acanthococcus devoniensis (Green), see *Eriococcus devoniensis* (Green)

Acantholyda erythrocephala (L.)	steel-blue sawfly
Acanthoscelides obtectus (Say)	dried bean beetle American seed beetle
ACANTHOSOMIDAE	shield bugs

see also CYDNIDAE, PENTATOMIDAE and SCUTELLERIDAE

Acarapis externus Morgenthaler	external bee mite
Acarapis woodi (Rennie)	acarine disease mite Isle of Wight disease mite
ACARI ACARINA	mites and ticks
Acarus siro L.	flour mite
see also *Thyreophagus entomophagus*	grain mite

Acaudus, subgenus of *Brachycaudus*
   Acedes pallescentella (Stainton), see *Tinea pallescentella*
   Aceria essigi (Hassan), see *Acalitus essigi*

Aceria ficus (Cotte)	fig mite

Aceria gracilis (Nalepa), see *Phyllocoptes gracilis*

Aceria lycopersici (Wolffenstein)	tomato erineum mite
Aceria macrochelus (Nalepa)	maple leaf solitary-gall mite

Aceria phloeocoptes (Nalepa), see *Acalitus phloeocoptes*
Aceria rudis (Canestrini), see *Acalitus rudis*
Aceria triradiatus (Nalepa), see *Eriophyes triradiatus*

Aceria tristriatus (Nalepa)	walnut leaf gall mite Persian walnut leaf blister mite
Acherontia atropos (L.)	death's head hawk moth
Acheta domesticus (L.)	house cricket

Achorutes (in part), see *Hypogastrura*
Achorutes armatus Nicolet, see *Hypogastrura armata*

Achroia grisella (Fabricius)	lesser wax moth

Acidia heraclei (L.), see *Euleia heraclei*

Acleris comariana (Leinig & Zeller)	strawberry tortrix (moth)

note: the name *comariana* was wrongly used in Pierce & Metcalfe, 1922, to describe *Acleris laterana* (Fabricius)
Acleris contaminana (Hübner), see *Acleris rhombana*

SCIENTIFIC NAMES—Arthropods

Scientific Name | Common Name

Acleris laterana (Fabricius) .. .. .. broad-barred button moth
   Acleris latifasciana (Haworth), see *Acleris laterana*
Acleris rhombana (Denis & Schiffermüller)   fruit tree tortrix (moth)
   see also species of *Apotomis, Archips,*   rhomboid tortrix
   Ditula, Hedya, Olethreutes and
   Pandemis
Acleris schalleriana (L.) .. .. .. viburnum tortrix (moth)
Acleris variegana (Denis & Schiffermüller)   garden rose tortrix (moth)
Aclypea opaca (L.) .. .. beet carrion beetle
ACRIDIDAE .. .. grasshoppers and locusts
                                             short-horned grasshoppers
   Acroclita naevana (Hübner), see *Rhopobota naevana*
   Acrolepia assectella (Zeller), see *Acrolepiopsis assectella*
   Acrolepiopsis assectella (Zeller): misspelling, see *Acrolepiopsis assectella*
Acrolepiopsis assectella (Zeller) .. .. leek moth
Acronicta psi (L.) .. .. grey dagger moth
                                               grey trident
Acronicta rumicis (L.) .. .. .. knotgrass moth
   Acronycta, see *Acronicta*
Aculops acericola (Nalepa) .. .. .. sycamore gall mite
Aculops lycopersici (Massee) .. .. tomato russet mite
Aculus fockeui (Nalepa & Trouessart) .. plum rust mite
                                             plum leaf mite (see also *Acalitus*
                                                  phloeocoptes)
   Aculus lycopersici (Massee), see *Aculops lycopersici*
Aculus schlechtendali (Nalepa) .. .. apple rust mite
                                             apple leaf and bud mite
Acyrthosiphon malvae (Mosley) .. .. pelargonium aphid
   Acyrthosiphon onobrychis (Boyer de Fonscolombe), see *Acyrthosiphon pisum*
   Acyrthosiphon pelargonii (Kaltenbach), see *Acyrthosiphon malvae*
   Acyrthosiphon pisi (Kaltenbach), see *Acyrthosiphon pisum*
Acyrthosiphon pisum (Harris) .. .. pea aphid
   Adalia 2-punctata (L.), see *Adalia bipunctata*
   Adalia 10-punctata (L.), see *Adalia decempunctata*
†*Adalia bipunctata* (L.) .. .. .. two-spot ladybird (beetle)
†*Adalia decempunctata* (L.) .. .. .. ten-spot ladybird (beetle)
Adelges abietis (L.) .. .. spruce pineapple-gall adelges
   see also *Adelges viridis*
   Adelges coccineus (Ratzeburg), see *Adelges laricis* Vallot
Adelges cooleyi (Gillette) .. .. .. Douglas fir adelges
   Adelges gallarumabietis (Degeer), see *Adelges abietis*
   Adelges geniculatus (Ratzeburg), see *Adelges laricis* Vallot
    note: the name *geniculatus* has been wrongly used in several published works to describe
       Adelges viridis (Ratzeburg)
   Adelges laricis (Hartig), see *Adelges viridis*
Adelges laricis Vallot .. .. larch adelges
                                         larch woolly aphid
Adelges nordmannianae (Eckstein) .. .. silver fir migratory adelges
   Adelges schneideri (Börner): *schneideri* (Börner) is a species synonym, but the name is
      retained for a subspecies, see *Adelges nordmannianae*
   Adelges nuesslini (Börner), see *Adelges nordmannianae*
Adelges piceae (Ratzeburg) .. .. .. silver fir adelges
   Adelges piceae canadensis Merker & Eichhorn, see *Adelges piceae*
   Adelges strobilobius (Kaltenbach), see *Adelges laricis* Vallot
Adelges viridana (Cholodkovsky) .. .. Japanese-larch adelges
Adelges viridis (Ratzeburg) .. .. .. spruce pineapple-gall adelges
   see also *Adelges abietis*
ADELGIDAE .. .. adelges
                                         conifer woolly aphids

SCIENTIFIC NAMES—Arthropods

| Scientific Name | Common Name |

Adoxophyes orana (Fischer von
   Röslerstamm) .. ..   summer fruit tortrix (moth)
Aedes caspius (Pallas) ⎫
Aedes detritus (Haliday) ⎭ .. .. ..   salt marsh mosquitoes
  Aegeria myopaeformis (Borkhausen), see *Synanthedon myopaeformis*
  Aegeria tipuliformis (Clerck), see *Synanthedon tipuliformis*
Aelia acuminata (L.) .. ..   bishop's mitre (bug)
Aeolothrips tenuicornis Bagnall .. ..   banded-wing flower thrips
Aglossa caprealis (Hübner) .. .. ..   small tabby moth
                              stable tabby moth
                              larva = murky meal caterpillar
Agonopterix conterminella (Zeller) .. ..   willow shoot moth
                              sallow flat-body moth
Agonopterix nervosa (Haworth) .. ..   carrot and parsnip flat-body moth
Agriopis aurantiaria (Hübner) .. ..   scarce umber moth
Agriopis marginaria (Fabricius) .. ..   dotted border moth
Agriotes lineatus (L.) ⎫
Agriotes obscurus (L.) ⎬ .. .. ..   common click beetles
Agriotes sputator (L.) ⎭   skipjacks
  see also *Athous* spp.   *larvae* = wireworms
Agromyza ambigua Fallén ⎫
Agromyza nigrella Rondani ⎭ .. ..   (fly) *larvae* = cereal leaf miners
  see also *Hydrellia* spp. and *Phytomyza nigra*
  Agromyza simplex Loew, see *Ophiomyia simplex*
Agrotis spp. .. ..   dart moths
  see also *Euxoa* spp.   *larvae* = cutworms
                         surface caterpillars ⎫ see also
                                    ⎭ *Noctua*
                                        pronuba
Agrotis exclamationis (L.) .. .. ..   heart and dart moth
Agrotis ipsilon (Hufnagel) .. .. ..   dark sword-grass moth
                              larva = black cutworm
  Agrotis nigricans (L.), see *Euxoa nigricans*
  Agrotis ravida (Denis & Schiffermüller), see *Spaelotis ravida*
  Agrotis saucia (Hübner), see *Peridroma saucia*
Agrotis segetum (Denis & Schiffermüller) ..   turnip moth
                              larva = common cutworm
Agrotis vestigialis (Hufnagel) .. ..   archer's dart moth
  Agrotis ypsilon (Hufnagel): misspelling, see *Agrotis ipsilon*
Aguriahana stellulata (Burmeister) .. ..   cherry leafhopper
Ahasverus advena (Waltl) .. .. ..   foreign grain beetle
†*Aleochara* spp. .. ..   small-headed rove beetles
†*Aleochara bilineata* Gyllenhal ⎫
†*Aleochara bipustulata* (L.) ⎭ .. ..   egg-eating rove beetles
  Aleurobius farinae (Degeer), see *Acarus siro*
  Aleurodes, see *Aleyrodes*
  Aleurodes ribium Douglas, see *Asterobemisia carpini*
  Aleurodes rubicola Douglas, see *Asterobemisia carpini*
  ALEURODIDAE, see ALEYRODIDAE
  Aleurotrachelus jelinekii (von Frauenfeld), see *Aleurotuba jelinekii*
Aleurotuba jelinekii (von Frauenfeld) ..   viburnum whitefly
  Aleyrodes azaleae Baker & Moles, see *Pealius azaleae*
  Aleyrodes brassicae Walker, see *Aleyrodes proletella*
  Aleyrodes fragariae Walker, see *Aleyrodes lonicerae*
Aleyrodes lonicerae Walker .. .. ..   honeysuckle whitefly
                              strawberry whitefly
Aleyrodes proletella (L.) .. .. ..   cabbage whitefly
ALEYRODIDAE .. ..   whiteflies
Allantus cinctus (L.) .. ..   banded rose sawfly
†*Allothrombium fuliginosum* (Hermann) ..   red velvet mite

SCIENTIFIC NAMES—Arthropods

Scientific Name	Common Name

Alnetoidia alneti (Dahlbom) .. .. .. fruit tree leaf hopper
   see also species of *Edwardsiana, Ribautiana, Typhlocyba* and *Zygina*
   Alnetoidia coryli (Tollin): *coryli* (Tollin) is a species synonym, but the name is retained for
     a subspecies, see *Alnetoidia alneti*
Alphitobius diaperinus (Panzer) ⎫ .. .. lesser mealworm beetles
Alphitobius laevigatus (Fabricius) ⎭   black fungus beetles
                                    larvae = lesser mealworms
   Alphitobius piceus: not *piceus* (Olivier, 1792), although misidentified as such in several later
     published works, see *Alphitobius diaperinus*
Alphitophagus bifasciatus (Say) .. .. two-banded fungus beetle
                                      waste grain beetle
Alsophila aescularia (Denis & Schiffermüller) March moth
Altica ericeti (Allard) .. .. heather flea beetle
Altica lythri (Aubé) .. .. large blue flea beetle
ALTICINAE .. .. flea beetles
   (in particular *Phyllotreta* spp.)
Alucita hexadactyla (L.) .. .. .. many plumed moth
                                      twenty plumed moth
   Amathes c-nigrum (L.), see *Xestia c-nigrum*
Amauromyza flavifrons (Meigen) .. .. (fly) *larva* = carnation leaf blotch miner
   Amaurosoma spp., see *Nanna* spp.
AMBLYCERA .. .. chewing lice
   see also ISCHNOCERA   biting lice (see also
                                      RHYNCHOPHTHIRINA)
†*Amblyseius finlandicus* (Oudemans) .. fruit tree red spider mite predator (mite)
   see also *Typhlodromus pyri*
   Amelia: not *Amelia* Hübner, although misidentified as such in Fernald, 1908, and in several
     later published works, see *Aphelia*
   Ametastegia carpini (Hartig), see *Protemphytus carpini*
Ametastegia glabrata (Fallén) .. .. dock sawfly
   Ametastegia pallipes (Spinola), see *Protemphytus pallipes*
   Amphidasis betularia (L.), see *Biston betularia*
Amphimallon solstitialis (L.) .. .. summer chafer (beetle)
   Amphimallus, see *Amphimallon*
Amphipyra tragopoginis (Clerck) .. .. mouse moth
   Amphorophora cosmopolitana Mason (in part), see *Hyperomyzus lactucae*
Amphorophora idaei (Börner) .. .. large raspberry aphid
Amphorophora rubi (Kaltenbach) .. .. bramble aphid
                                      rubus aphid
Anagasta, subgenus of *Ephestia*
   Anagasta kuehniella Zeller, see *Ephestia kuehniella*
Anaglyptus mysticus (L.) .. .. .. grey-coated longhorn beetle
ANALGOIDEA .. .. feather mites
Anaphothrips obscurus (Müller) .. .. striate thrips
                                      American grass thrips
Anaphothrips orchidaceus Bagnall .. .. yellow orchid thrips
                                      yellow thrips
   Anaphothrips striatus (Osborn), see *Anaphothrips obscurus*
**Anarsia lineatella* Zeller .. .. .. (moth) *larva* = peach twig borer
   Anaticola anatis (Fabricius), see *Anaticola crassicornis*
Anaticola crassicornis (Scopoli) .. .. duck wing louse
                                      slender duck louse
   Anaticola squalidus (Nitzsch), see *Anaticola crassicornis*
†*Anatis ocellata* (L.) .. .. pine ladybird (beetle)
                                      eyed ladybird
Andrena spp. .. .. andrenas
                                      burrowing bees
                                        solitary bees

SCIENTIFIC NAMES—Arthropods

Scientific Name	Common Name
Andrena armata (Gmelin in Linnaeus), see *Andrena fulva*	
Andrena fulva (Müller in Allioni)	tawny burrowing bee
Andricus spp.	oak gall cynipids
	oak gall wasps
Andricus albopunctatus (Schlechtendal)	oak leaf gall cynipid (wasp)
see also *Andricus quadrilineatus* and *Andricus solitarius*	
Andricus circulans Mayr, see *Andricus kollari*	
Andricus curvator Hartig	oak bud collared-gall cynipid (wasp)
Andricus fecundator (Hartig)	larch cone gall cynipid (wasp)
	artichoke gall wasp
	hop gall wasp
Andricus furunculus (Beyerinck), see *Andricus ostreus*	
Andricus inflator Hartig	oak bud globular-gall cynipid (wasp)
Andricus kollari (Hartig)	marble gall wasp
Andricus nudus Adler	oak catkin gall cynipid (wasp)
Andricus ostreus (Hartig)	oak leaf oyster-gall cynipid (wasp)
Andricus quadrilineatus Hartig	oak leaf gall cynipid (wasp)
see also *Andricus albopunctatus* and *Andricus solitarius*	
Andricus quercuscalicis (Burgsdorf)	acorn cup gall cynipid (wasp)
Andricus quercusradicis (Fabricius)	oak root truffle-gall cynipid (wasp)
Andricus solitarius (Fonscolombe)	oak leaf gall cynipid (wasp)
see also *Andricus albopunctatus* and *Andricus quadrilineatus*	
Andricus testaceipes Hartig	oak red barnacle-gall cynipid (wasp)
Androlaelaps casalis (Berlese)	poultry litter mite
	cosmopolitan nest mite
Androniscus dentiger Verhoeff	cellar woodlouse
Aneuloboiulus, subgenus of *Cylindroiulus*	
Angitia spp., see *Diadegma* spp.	
Anisandrus dispar (Fabricius), see *Xyleborus dispar*	
Anisandrus dryographus (Ratzeburg), see *Xyleborus dryographus*	
Anisandrus saxeseni (Ratzeburg), see *Xyleborus saxeseni*	
ANISOPODIDAE	{ sewage filter-bed flies
(in particular *Sylvicola* spp.)	{ window gnats
Anisopteryx aescularia (Denis & Schiffermüller), see *Alsophila aescularia*	
Anisopus spp., see *Sylvicola* spp.	
ANOBIIDAE	furniture beetles
	larvae = woodworms
Anobium domesticum (Fourcroy), see *Anobium punctatum*	
Anobium punctatum (Degeer)	common furniture beetle
	furniture beetle
	larva = woodworm
Anobium striatum: not *striatum* Fabricius, 1787, although misidentified as such in several later published works, see *Anobium punctatum*	
Anoecia corni (Fabricius)	dogwood—grass aphid
	dogwood aphid
Anomoea permunda (Harris), see *Anomoia permunda*	
Anomoia permunda (Harris)	barberry seed fly
	barberry fly
	berberis seed fly
Anoncodes melanura (L.), see *Nacerdes melanura*	
Anopheles atroparvus van Thiel	common malarial anopheles mosquito
	common anopheles mosquito
Anopheles maculipennis: not *maculipennis* Meigen, 1818, although misidentified as such in several later published works, see *Anopheles atroparvus*	
note: *Anopheles maculipennis* Meigen does not occur in Britain	
Anopheles messeae Falleroni	Messea's anopheles mosquito
Anopheles plumbeus Stephens	tree-hole mosquito
Anoplonyx destructor Benson	Benson's larch sawfly
ANOPLURA	sucking lice

SCIENTIFIC NAMES—Arthropods

Scientific Name	Common Name
†*Anotylus rugosus* (Fabricius)	wrinkled rove beetle
Anotylus tetracarinatus (Block)	fly-in-the-eye beetle
Anthicus floralis (L.)	narrow-necked harvest beetle
†*Anthocoris* spp.	anthocorid bugs
	flower bugs
†*Anthocoris confusus* Reuter	oak flower bug
†*Anthocoris gallarumulmi* (Degeer)	elm gall bug
†*Anthocoris nemorum* (L.)	common flower bug

Anthonomus cinctus Redtenbacher, see *Anthonomus piri*

Anthonomus piri Kollar	apple bud weevil
Anthonomus pomorum (L.)	apple blossom weevil

Anthonomus pyri: misspelling, see *Anthonomus piri*

Anthonomus rubi (Herbst)	strawberry blossom weevil
	elephant beetle
	needle-bug

Anthophila pariana (Clerck), see *Choreutis pariana*

Anthrenocerus australis (Hope)	Australian carpet beetle
Anthrenus spp.	carpet beetles
	larvae = woolly bears
**Anthrenus flavipes* Le Conte	furniture carpet beetle
Anthrenus museorum (L.)	museum beetle
Anthrenus sarnicus Mroczkowski	Guernsey carpet beetle
Anthrenus verbasci (L.)	varied carpet beetle

Anuraphis amygdali (Buckton), see *Brachycaudus schwartzi*
Anuraphis cardui (L.), see *Brachycaudus cardui*
Anuraphis crataegi: not *crataegi* (Kaltenbach, 1843), although misidentified as such (in part) in Theobald 1927, and in several later published works, see *Dyasphis devecta*

Anuraphis farfarae (Koch)	pear—coltsfoot aphid

Anuraphis helichrysi (Kaltenbach), see *Brachycaudus helichrysi*
Anuraphis masseei (Theobald), see *Brachycaudus persicae*
Anuraphis persicae-niger (Smith), see *Brachycaudus persicae*
Anuraphis pyri (Boyer de Fonscolombe), see *Dysaphis pyri*
Anuraphis roseus (Baker), see *Dysaphis plantaginea*

Anuraphis subterranea (Walker)	pear—parsnip aphid

Anuraphis tulipae (Boyer de Fonscolombe), see *Dysaphis tulipae*

Anurida granaria (Nicolet)	plump white springtail
†*Anystis* spp.	whirligig mites
	anystis mites
**Aonidiella aurantii* (Maskell)	California red scale

Apamea basilinea (Denis & Schiffermüller), see *Apamea sordens*
Apamea secalis (L.), see *Mesapamea secalis*

Apamea sordens (Hufnagel)	rustic shoulder knot moth
	larva = wheat cutworm
†*Apanteles* spp.	apanteles (parasitic wasps)
†*Apanteles glomeratus* (L.)	common apanteles (parasitic wasp)

Apate capucina Fabricius, see *Xylopsocus capucinus*

*Apate monachus Fabricius	(beetle) *larva* = black borer
*Apate terebrans (Pallas)	(beetle) *larva* = African gimlet

Apatele, see *Acronicta*
Apatele psi (L.), see *Acronicta psi*

Aphelia paleana (Hübner)	timothy tortrix (moth)
	plain yellow twist
Aphelia viburnana (Denis & Schiffermüller)	bilberry tortrix (moth)
†*Aphelinus* spp.	aphid parasitic wasps

see also *Aphidius* spp.

†*Aphelinus mali* (Haldeman)	woolly aphid parasite (wasp)
	woolly aphis parasite
†*Aphidecta obliterata* (L.)	conifer ladybird (beetle)

see also *Exochomus quadripustulatus*

SCIENTIFIC NAMES—Arthropods

Scientific Name	Common Name
APHIDIDAE	aphids greenflies plant-lice

†*Aphidius* spp. .. .. aphid parasitic wasps
    see also *Aphelinus* spp.
    Aphidius avenae Haliday, see *Aphidius picipes*
    Aphidius brassicae (Marshall in André), see *Diaeretiella rapae*
†*Aphidius matricariae* Haliday ⎫
†*Aphidius picipes* (Nees) ⎭ .. .. peach—potato aphid parasitic wasps
    Aphis abietina Walker, see *Elatobium abietinum*
    Aphis avenae: not *avenae* Fabricius, 1775, although misidentified as such in several later published works, see *Rhopalosiphum padi*
    Aphis carbocolor Gillette, see *Aphis rumicis*
Aphis craccivora Koch .. .. .. cowpea aphid
                                                   black legume aphid
Aphis fabae Scopoli .. .. black bean aphid
                                                   black dolphin
                                                   blackfly
Aphis farinosa Gmelin .. .. .. small willow aphid
Aphis gossypii Glover .. .. .. melon and cotton aphid
Aphis grossulariae Kaltenbach .. .. gooseberry aphid
Aphis idaei van der Goot .. .. .. small raspberry aphid
                                                   raspberry aphid
Aphis lambersi (Börner) .. .. .. permanent carrot aphid
Aphis nasturtii Kaltenbach .. .. buckthorn—potato aphid
Aphis nerii Boyer de Fonscolombe .. oleander aphid
Aphis plantaginis Goeze .. .. .. plantain aphid
Aphis pomi Degeer .. .. green apple aphid
                                                 permanent apple aphid
    Aphis rhamni: not *rhamni* Boyer de Fonscolombe, 1841, although misidentified as such in several later published works, see *Aphis nasturtii*
Aphis ruborum (Börner) .. .. .. permanent blackberry aphid
Aphis rumicis L. .. .. permanent dock aphid
    Aphis saliceti Kaltenbach, see *Aphis farinosa*
Aphis sambuci L. .. .. elder aphid
Aphis schneideri (Börner) .. .. permanent currant aphid
Aphis triglochinis Theobald .. .. red currant—arrowgrass aphid
    Aphis varians: not *varians* Patch, 1914, although misidentified as such in several later published works, see *Aphis schneideri*
Aphis viburni Scopoli .. .. .. viburnum aphid
Aphodius spp. .. .. dung beetles
    Aphomia gularis (Zeller), see *Paralipsa gularis*
Aphomia sociella (L.) .. .. .. bumble-bee wax moth
                                               bee moth
                                               green honey moth
Aphrodes bicinctus (Schrank) .. .. strawberry leafhopper
Aphrophora alni (Fallén) .. .. .. alder froghopper
Aphrophora salicina (Goeze) .. .. willow froghopper
    Aphrophora salicis (Degeer), see *Aphrophora salicina*
    Aphrophora spumaria: not *spumaria* (Linnaeus, 1785), although misidentified as such in several later published works, see *Aphrophora alni*
Aphthona euphorbiae (Schrank) .. .. large flax flea beetle
†*Aphytis mytilaspidis* (LeBaron) .. .. scale insect parasite (wasp)
Apion spp. .. .. flower weevils
    Apion aestivum Germar, see *Apion trifolii*
Apion apricans Herbst ⎫
Apion assimile Kirby ⎭ .. .. .. clover seed weevils
                                               red clover seed weevils
    see also *Apion trifolii*
Apion dichroum Bedel .. .. .. white clover seed weevil

SCIENTIFIC NAMES—Arthropods

Scientific Name	Common Name

Apion flavipes: not *flavipes* (Fabricius, 1775), although misidentified as such in Paykull, 1792, and in several later published works, see *Apion dichroum*
Apion frumentarium (Paykull), see *Apion haematodes*
   Apion haematodes Kirby .. .. .. sheep's sorrel gall weevil
   Apion pomonae (Fabricius) .. .. .. vetch seed weevil
                                                         tare seed weevil
   Apion radiolus (Marsham) .. .. .. hollyhock weevil
   Apion trifolii (L.) .. .. clover seed weevil
      see also *Apion apricans* and *Apion*   red clover seed weevil
      assimile
   Apion vorax Herbst .. .. .. bean flower weevil
†*Apis mellifera* L. .. .. honey bee
Apocheima pilosaria (Denis & Schiffermüller) pale brindled beauty moth
Apocremnus, subgenus of *Psallus*
Apoderus coryli (L.) .. .. (weevil) *larva* = hazel leaf roller
Apotomis spp. .. .. fruit tree tortrices (moths)
   see also species of *Acleris, Archips, Ditula, Hedya, Olethreutes* and *Pandemis*
      Apotomis pruniana (Hübner), see *Hedya pruniana*
Appelia, subgenus of *Brachycaudus*
   Appelia schwartzi Börner, see *Brachycaudus schwartzi*
   Apterococcus fraxini (Newstead), see *Pseudochermes fraxini*
   Aptinothrips lubbocki (Bagnall), see *Aptinothrips rufus*
   Aptinothrips rufus (Gmelin), see *Aptinothrips rufus* (Haliday)
   Aptinothrips rufus (Haliday) ⎫
   Aptinothrips stylifer Trybom ⎬ .. .. grass thrips
†*Aptus mirmicoides* (Costa) .. .. .. ant damsel bug
ARACHNIDA .. .. spiders, mites, ticks, harvestmen, etc.
ARADIDAE .. .. flat bugs
                                                    bark bugs
Aradus cinnamomeus (Panzer) .. .. pine flat bug
Araecerus fasciculatus (Degeer) .. .. coffee bean weevil
                                                    cacao weevil
ARANEAE .. .. spiders
Archaeopsylla erinacei (Bouché) .. .. hedgehog flea
Archiboreoiulus pallidus (Brade-Birks) .. snake millepede
   see also *Boreoiulus tenuis, Blaniulus guttulatus, Cylindroiulus londinensis* and
      Tachypodoiulus niger
Archippus, subgenus of *Archips*
Archips crataegana (Hübner) .. .. fruit tree tortrix (moth)
   see also species of *Acleris, Apotomis,*   brown oak tortrix (see also *Archips*
      Ditula, Hedya, Olethreutes and   *xylosteana*)
      Pandemis
Archips podana (Scopoli) .. .. .. large fruit tree tortrix (moth)
                                                    fruit tree tortrix (see also species of *Acleris,*
                                                         Apotomis, Ditula, Hedya, Olethreutes
                                                         and *Pandemis*)
Archips oporana: not *oporana* (Linnaeus, 1758) although at least two different species have
   been misidentified as such in several later published works, see *Archips crataegana* and
      Archips podana
Archips roborana (Hübner), see *Archips crataegana*
Archips rosana (L.) .. .. rose tortrix (moth)
                                                    rose twist
Archips xylosteana (L.) .. .. .. variegated golden tortrix (moth)
                                                    brown oak tortrix (see also *Archips*
                                                         crataegana)
                                                    forked red-barred twist
*Argas persicus (Oken) .. .. .. fowl tick
Argas reflexus (Fabricius) .. .. .. pigeon soft tick
                                                    Canterbury tick

SCIENTIFIC NAMES—Arthropods

| Scientific Name | Common Name |

Argas vespertilionis (Latreille) .. .. bat soft tick
                                                  Blyborough tick
ARGASIDAE .. .. soft ticks
Arge ochropus (Gmelin in Linnaeus) .. large rose sawfly
   Argyresthia atmoriella Bankes, see *Argyresthia laevigatella*
Argyresthia conjugella Zeller .. .. apple fruit moth
                                         larva = apple fruit miner
   Argyresthia curvella: not *curvella* (Linnaeus, 1761), although misidentified as such in several later published works, see *Argyresthia pruniella*
Argyresthia pruniella (Clerck) .. .. cherry fruit moth
Argyresthia laevigatella Herrich-Schäffer larch shoot moth
                                         larva = larch shoot borer
   Argyresthia nitidella: not *nitidella* (Fabricius, 1787), although misidentified as such in several later published works, see *Argyresthia pruniella*
   Argyroploce: a generic synonym, but the name is retained for a subgenus of *Olethreutes*, see species of *Apotomis*, *Hedya* and *Olethreutes*
   Argyroploce nubiferana (Haworth), see *Hedya dimidioalba*
   Argyroploce pruniana (Hübner), see *Hedya pruniana*
   Argyroploce variegana (Hübner), see *Hedya dimidioalba*
Argyrotaenia ljungiana (Thunberg) .. .. grey red-barred tortrix (moth)
   Argyrotaenia pulchellana (Haworth), see *Argyrotaenia ljungiana*
   Argyrotoza comariana (Zeller), see *Acleris comariana*
Arhopalus rusticus (L.) .. .. .. rusty longhorn beetle
Armadillidium depressum Brandt .. .. western stout pillbug (woodlouse)
Armadillidium nasatum Budde-Lund .. blunt snout pillbug (woodlouse)
Armadillidium vulgare (Latreille) .. .. common pillbug (woodlouse)
Arnoldiola quercus (Binnie) .. .. oak terminal-shoot gall midge
Aromia moschata (L.) .. .. .. musk beetle
Artacris macrorhynchus (Nalepa) .. .. maple bead-gall mite
   Artogeia rapae (L.), see *Pieris rapae*
Asemum striatum (L.) .. .. .. pine longhorn beetle
ASILIDAE .. .. robber flies
Asiomorpha coarctata (Saussure) .. .. hothouse millepede
   Asiphum tremulae (L.), see *Pachypappa tremulae*
Asphondylia sarothamni Loew .. .. broom gall midge
Aspidapion, subgenus of *Apion*
   Aspidiotus britannicus Newstead, see *Dynaspidiotus britannicus*
   Aspidiotus hederae: not *hederae* (Vallot, 1829), although misidentified as such in several later published works, see *Aspidiotus nerii*
Aspidiotus nerii Bouché .. .. .. oleander scale
   Aspidiotus ostreaeformis Curtis, see *Quadraspidiotus ostreaeformis*
   Aspidiotus perniciosus Comstock, see *Comstockaspis perniciosa*
   Asterobemisia avellanae (Signoret), see *Asterobemisia carpini*
Asterobemisia carpini (Koch) .. .. hornbeam whitefly
   Asterochiton carpini (Koch), see *Asterobemisia carpini*
Asterodiaspis minus (Lindinger) ⎫
Asterodiaspis quercicola (Bouché) ⎬ .. oak pit scales (Homoptera)
Asterodiaspis variolosa (Ratzeburg) ⎭   oak pit gallers
   Athalia colibri (Christ), see *Athalia rosae*
Athalia rosae (L.) .. .. turnip sawfly
Atherix spp. .. .. snipe flies
   see also *Symphoromyia* spp.
Atheta spp. .. .. small rove beetles
Athous spp. .. .. click beetles
   see also species of *Agriotes*   skipjacks
                                           larvae = wireworms
Athous haemorrhoidalis (Fabricius) .. garden click beetle
                                         larva = garden wireworm
Atomaria linearis Stephens .. .. pygmy mangold beetle
                                         pygmy beetle

SCIENTIFIC NAMES—Arthropods

Scientific Name	Common Name

†*Atractotomus mali* (Meyer-Duer) .. .. black apple capsid (bug)
    Attagenus megatoma (Fabricius), see *Attagenus unicolor*
Attagenus pellio (L.) .. .. fur beetle
                                          two-spot carpet beetle
    Attagenus piceus (Olivier, 1790), not *piceus* (Thunberg, 1781), see *Attagenus unicolor*
Attagenus unicolor (Brahm) .. .. .. black carpet beetle
Attelabus nitens (Scopoli) .. .. .. oak leaf roller weevil
Aulacaspis rosae (Bouché) .. .. .. rose scale
                                          scurfy scale
Aulacorthum circumflexum (Buckton) .. mottled arum aphid
Aulacorthum solani (Kaltenbach) .. .. glasshouse and potato aphid
                                          foxglove aphid
    Austrotortrix postvittana (Walker), see *Epiphyas postvittana*
Autographa gamma (L.) .. .. .. silver y moth
                                          larva = beetworm

Aylax minor Hartig ⎫
Aylax papaveris (Perris) ⎬ .. .. .. poppy gall cynipids (wasps)

    Balaninus nucum (L.), see *Curculio nucum*
    Balanobius salicivorus (Paykull), see *Curculio salicivorus*
    Balioptera tripunctata (Fallén), see *Geomyza tripunctata*
    Baliothrips graminum (Uzel), see *Stenothrips graminum* Uzel
    Barathra albidilinea (Haworth), see *Mamestra brassicae*
    Barathra brassicae (L.) see *Mamestra brassicae*
Barynotus obscurus (Fabricius) .. .. ground weevil
Barypeithes spp. .. .. broad-nosed weevils
Barypeithes araneiformis (Schrank) .. smooth broad-nosed weevil
                                          strawberry fruit weevil (see *Barypeithes pellucidus*)
Barypeithes pellucidus (Boheman) .. .. hairy broad-nosed weevil
                                          strawberry fruit weevil (see *Barypeithes araneiformis*)
    Barypithes: misspelling, see *Barypeithes*
†*Basalys tritoma* Thomson .. .. .. frit fly parasite (wasp)
    Batodes angustiorana (Haworth), see *Ditula angustiorana*
Batophila aerata (Marsham) ⎫
Batophila rubi (Paykull) ⎬ .. .. raspberry flea beetles
    Bdellonyssus bacoti (Hirst), see *Ornithonyssus bacoti*
    Bdellonyssus sylviarum (Canestrini & Fanzago), see *Ornithonyssus sylviarum*
†*Bembidion* spp. .. .. brassy ground beetles
*Bemisia tabaci (Gennadius) .. .. .. tobacco whitefly
                                          cotton whitefly
                                          sweet potato whitefly
Bibio hortulanus (L.) .. .. March fly
Bibio marci (L.) .. .. St. Mark's fly
BIBIONIDAE .. .. St. Mark's flies
    Biorhiza aptera (Fabricius), see *Biorhiza pallida*
Biorhiza pallida (Olivier) .. .. .. oak-apple gall wasp
    Biorhiza quercusterminalis (Fabricius), see *Biorhiza pallida*
Biston betularia (L.) .. .. peppered moth
                                          pepper and salt moth
                                          larva = hop-cat
    Biston hirtaria (Clerck), see *Lycia hirtaria*
    Blaniulus guttulatus (Bosc): attribution to author incorrect, see *Blaniulus guttulatus* (Fabricius)

SCIENTIFIC NAMES—Arthropods

Scientific Name   **Common Name**

Blaniulus guttulatus (Fabricius)   ..   ..   spotted snake millepede
                                                      snake millepede (see also *Archiboreoiulus pallidus, Boreoiulus tenuis, Cylindroiulus londinensis* and *Tachypodoiulus niger*)
                                                    spotted millepede
Blaps spp.   ..   ..   cellar beetles
Blastesthia turionella (L.)   ..   ..   ..   pine bud moth
Blasticotoma filiceti Klug   ..   ..   ..   (sawfly) *larva* = fern stem borer
Blastobasis decolorella (Wollaston) ..   straw-coloured apple moth
                                                     larva = apple skin spoiler
     Blastodacna atra (Haworth), see *Spuleria atra* (Haworth)
     Blastophagus piniperda (L.), see *Tomicus piniperda*
     Blastotere, subgenus of *Argyresthia*
         Blastotere laevigatella (Herrich-Schäffer), see *Argyresthia laevigatella*
Blatta orientalis L.   ..   ..   common cockroach
                                                    blackbeetle
                                                   oriental cockroach
Blattella germanica (L.)   ..   ..   ..   German cockroach
                                                    croton-bug
                                                    shiner
                                                    steamfly
     BLATTODEA   ..   ..   cockroaches
     Blennocampa geniculata (Hartig), see *Monophadnoides geniculatus*
Blennocampa pusilla (Klug)   ..   ..   ..   leaf-rolling rose sawfly
†*Blepharidopterus angulatus* (Fallén)   ..   black-kneed capsid (bug)
     Blitophaga opaca (L.), see *Aclypea opaca*
†*Bombus* spp.   ..   ..   bumble-bees
                                                    humble-bees
     BOMBYLIIDAE   ..   ..   bee flies
*†*Bombyx mori* (L.)   ..   ..   common silk moth
                                                    larva = mulberry silkworm
                                                    silkworm
     BORBORIDAE, see SPHAEROCERIDAE
Boreoiulus tenuis (Bigler)   ..   ..   ..   snake millepede
     see also *Archiboreoiulus pallidus, Blaniulus guttulatus, Cylindroiulus londinensis* and *Tachypodoiulus niger*
     Borkhausenia pseudospretella (Stainton), see *Hofmannophila pseudospretella*
     BOSTRICHIDAE   ..   ..   ..   (beetle) *larvae* = wood borers
**Bostrychoplites cornutus* (Olivier)   ..   ..   (beetle) *larva* = African horned wood borer
Botanophila gnava (Meigen) ..   ..   ..   lettuce seed fly
Bourletiella hortensis (Fitch)   ..   ..   garden springtail
     Bourletiella signatus (Nicolet), see *Bourletiella hortensis*
Bovicola bovis L.   ..   ..   cattle biting louse
Bovicola caprae (Gurlt)   ..   ..   ..   common goat biting louse
Bovicola ovis (Schrank)   ..   ..   ..   sheep biting louse
Bracheioiulus, subgenus of *Cylindroiulus*
Brachonyx pineti (Paykull)   ..   ..   ..   pine needle weevil
     Brachycaudus amygdali (Buckton), see *Brachycaudus schwartzi*
Brachycaudus cardui (L.)   ..   ..   ..   thistle aphid
Brachycaudus helichrysi (Kaltenbach)   ..   leaf-curling plum aphid
Brachycaudus persicae (Passerini)   ..   ..   black peach aphid
     Brachycaudus persicaecola (Bosiduval), see *Brachycaudus persicae*
     Brachycaudus prunicola: not *prunicola* (Kaltenbach, 1843), although misidentified as such in several later published works, see Brachycaudus persicae
Brachycaudus schwartzi (Börner)   ..   ..   peach aphid
Brachydesmus superus Latzel   ..   ..   flat millepede
     see also *Polydesmus angustus*   flat-backed millepede
Brachypterolus pulicarius (L.)   }
Brachypterolus vestitus (Kiesenwetter)}   ..   antirrhinum beetles

SCIENTIFIC NAMES—Arthropods

Scientific Name   **Common Name**

Brachyrhinus sulcatus (Fabricius), see *Otiorhynchus sulcatus*
BRACONIDAE .. .. braconids (wasps)
Bradysia spp. .. .. sciarid flies
Bradysia brunnipes (Meigen) .. .. mushroom sciarid (fly)
    see also species of *Lycoriella*
    Bradysia pectoralis: not *pectoralis* (Staeger, 1840), although misidentified as such in Edwards, 1925, and in several later published works, see *Bradysia tritici*
Bradysia tritici (Coquillett) .. .. .. moss fly
Braula coeca Nitsch .. .. bee louse
    Bregmatothrips iridis Watson, see *Frankliniella iridis*
Bremiola onobrychidis (Bremi) .. .. sainfoin leaf midge
Brevicoryne brassicae (L.) .. .. .. cabbage aphid
                                     mealy cabbage aphid
Brevipalpus spp. .. .. false spider mites
                                       false red spider mites
BRUCHIDAE .. .. pulse beetles
    Bruchophagus funebris (Howard), see *Bruchophagus gibbus*
Bruchophagus gibbus (Boheman) .. .. lucerne chalcid (wasp)
    Bruchus affinis: not *affinis* Froelich, 1799, although misidentified as such in several later published works, see *Bruchus rufimanus*
Bruchus spp. .. .. pea beetles, bean beetles
†*Bruchus ervi* Froelich .. .. Mediterranean pulse beetle
    Bruchus obtectus Say, see *Acanthoscelides obtectus*
Bruchus pisorum (L.) .. .. .. pea beetle
                                       pea seed beetle
Bruchus rufimanus Boheman .. .. bean beetle
                                       bean seed beetle
Bryobia cristata (Dugès) .. .. .. grass—pear bryobia (mite)
                                       pear bryobia
Bryobia kissophila van Eyndhoven .. .. ivy bryobia
                                       ivy mite
Bryobia praetiosa Koch .. .. .. clover bryobia
                                       clover mite
    note: the name *praetiosa* has been wrongly used variously in several published works to describe *Bryobia cristata, Bryobia kissophila, Bryobia ribis* and *Bryobia rubrioculus*
Bryobia ribis Thomas .. .. .. gooseberry bryobia
                                       gooseberry mite
                                       gooseberry red spider mite
Bryobia rubrioculus (Scheuten) .. .. apple and pear bryobia
                                       brown mite
Bupalus piniaria (L.) .. .. bordered white moth
                                       larva = pine looper
BUPRESTIDAE .. .. jewel beetles
                                       larvae = flat-headed borers
†*Buprestis aurulenta* (L.) .. .. .. golden buprestid (beetle)
Byctiscus betulae (L.) .. .. .. hazel leaf roller weevil
Byctiscus populi (L.) .. .. poplar leaf roller weevil
Byturus tomentosus (Degeer) .. .. raspberry beetle
                                       loganberry beetle

Cabera pusaria (L.) .. .. common white wave moth
    Cacoecia crataegana (Hübner), see *Archips crataegana*
    Cacoecia oporana: not *oporana* (Linnaeus, 1758), although misidentified as such in several later published works, see *Archips crataegana* and *Archips podana*
    Cacoecia podana (Scopoli), see *Archips podana*
    Cacoecia pronubana (Hübner), see *Cacoecimorpha pronubana*
    Cacoecia roborana (Hübner), see *Archips crataegana*

SCIENTIFIC NAMES—Arthropods

Scientific Name	Common Name

Cacoecimorpha pronubana (Hübner) .. carnation tortrix (moth)
    Mediterranean carnation leaf roller
    Cadra: a generic synonym, but the name is retained for a subgenus, see *Ephestia*
Caenorhinus, subgenus of *Rhynchites*
    Caenorhinus aequatus (L.), see *Rhynchites aequatus*
    Caenorhinus germanicus (Herbst), see *Rhynchites germanicus*
    Calandra: invalid name, see *Sitophilus*
    Caliroa aethiops (Fabricius), see *Endelomyia aethiops*
Caliroa annulipes (Klug) .. .. .. (sawfly) *larva* = oak slugworm
Caliroa cerasi (L.) .. .. pear slug sawfly
    pear and cherry sawfly
    larva = pear and cherry slugworm
    Caliroa limacina (Retzius), see *Caliroa cerasi*
Callaphis juglandis (Goeze) .. .. .. large walnut aphid
    Callidium variabilis (L.), see *Phymatodes testaceus*
Callidium violaceum (L.) .. .. .. violet tanbark beetle
    Callimorpha jacobaeae (L.), see *Tyria jacobaeae*
    Calliphora erythrocephala (Meigen), see *Calliphora vicina*
Calliphora vicina Robineau-Desvoidy ⎫
Calliphora vomitoria (L.) ⎬ .. bluebottles
    blow-flies
CALLIPHORIDAE .. .. bluebottles, greenbottles, flesh flies, etc.
    see also SARCOPHAGIDAE
Calliteara pudibunda (L.) .. .. .. pale tussock moth
    larva = hop-dog
Callosobruchus spp. .. .. cowpea beetles
Calocoris fulvomaculatus (Degeer) .. .. hop capsid
    needle-nosed hop bug
    shy bug
Calocoris norvegicus (Gmelin) .. .. potato capsid (bug)
Caloglyphus berlesei (Michael) .. .. wet grain mite
Caloptilia azaleella (Brants) .. .. .. (moth) *larva* = azalea leaf miner
Caloptilia syringella (Fabricius) .. .. (moth) *larva* = lilac leaf miner
†*Calosoma inquisitor* (L.) .. .. .. oakwood ground beetle
Camponotus spp. .. .. carpenter ants
    Campoplex oxyacanthae Boie, see *Dusona oxyacanthae*
    Campylochirus caviae (Hirst), see *Chirodiscoides caviae* Hirst
Campylomyza ormerodi (Kieffer) .. .. red clover gall gnat
    Capitophorus fragariae (Theobald), see *Chaetosiphon fragaefolii*
    Capitophorus galeopsidis (Kaltenbach), see *Cryptomyzus galeopsidis*
    Capitophorus ribis (L.), see *Cryptomyzus ribis*
CAPSIDAE, see MIRIDAE
    Capua reticulana (Hübner), see *Adoxophyes orana*
CARABIDAE .. .. tiger beetles and ground beetles
CARABINAE .. .. ground beetles
    see also HARPALINI
†*Carabus nemoralis* Müller .. .. .. forest ground beetle
†*Carabus violaceus* L. .. .. violet ground beetle
Curadrina clavipalpis (Scopoli) .. .. pale mottled willow moth
    Carausius morosus Brunner: attribution to author incorrect, see next entry
Carausius morosus (Sinéty) .. .. .. laboratory stick insect
    Carpocapsa pomonella (L.), see *Cydia pomonella*
Carpoglyphus lactis (L.) .. .. .. dried fruit mite
Carpophilus spp. .. .. dried fruit beetles
    Cartodere filum (Aubé), see *Dienerella filum*
Carulaspis juniperi (Bouché) ⎫
Carulaspis minima (Targioni-Tozzetti) ⎬ .. juniper scales
    Carulaspis visci: description of *visci* (Schrank, 1781) is unresolved; in later published works the name has been applied to two species: *Carulaspis juniperi* and *Carulaspis minima* (see above)

19

SCIENTIFIC NAMES—Arthropods

Scientific Name	Common Name
Caryedon fuscus (Goeze), see *Caryedon serratus*	
Caryedon gonagra (Fabricius), see *Caryedon serratus*	
**Caryedon serratus* (Olivier)	groundnut bruchid (beetle)
	larva = groundnut borer
Cassida spp.	tortoise beetles
Cathartus quadricollis (Guérin-Méneville)	square-necked grain beetle
Caulophilus latinasus: not *latinasus* (Say, 1831), although misidentified as such in several later published works, see *Caulophilus oryzae*	
Caulophilus oryzae (Gyllenhal)	broad-nosed grain weevil
Caulotrupis aeneopiceus (Boheman), see *Caulotrupodes aeneopiceus*	
Caulotrupodes aeneopiceus (Boheman)	timber borer (weevil)
Cavariella spp.	willow aphids
Cavariella aegopodii (Scopoli)	willow—carrot aphid
Cavariella theobaldi (Gillette & Bragg)	willow—parsnip aphid
Cecidomyia baeri (Prell), see *Contarinia baeri*	
CECIDOMYIIDAE	gall midges
	cecid midges
Cecidophyopsis psilaspis (Nalepa)	yew gall mite
	yew big bud mite
Cecidophyopsis ribis (Westwood)	black currant gall mite
	big bud mite
	currant bud mite
Celaena secalis (L.), see *Mesapamea secalis*	
Cenopalpus pulcher (Canestrini & Fanzago)	flat scarlet mite
Cephalcia alpina: not *alpina* (Klug, 1808), although misidentified as such in several later published works, see *Cephalcia lariciphila*	
Cephalcia lariciphila (Wachtl)	web-spinning larch sawfly
	larch webspinner
Cephenemyia auribarbis (Meigen)	deer nostril fly
CEPHIDAE	stem sawflies
Cephus pygmaeus: misspelling see *Cephus pygmeus*	
Cephus pygmeus (L.)	wheat stem sawfly
Cephus tabidus (Fabricius), see *Trachelus tabidus*	
CERAMBYCIDAE	longhorn beetles
	longicorn beetles
	timbermen
‡*Cerambyx cerdo* L.	large oak longhorn beetle
	capricorn beetle
Ceramica pisi (L.)	broom moth
	broom brocade moth
Cerapteryx graminis (L.)	antler moth
Cerataphis orchidearum (Westwood)	orchid aphid
	scale aphid
**Ceratitis capitata* (Wiedemann)	Mediterranean fruit fly
Ceratophyllus gallinae (Schrank)	European chicken flea
	hen flea
Ceratophysella, subgenus of *Hypogastrura*	
CERATOPOGONIDAE	biting midges
(in particular *Culicoides* spp.)	
CERCOPIDAE	froghoppers
	cuckoo-spit insects
Cercopis sanguinea (Geoffroy in Fourcroy), see *Cercopis vulnerata*	
Cercopis vulnerata Illiger in Rossi	red and black froghopper
Cerobasis guestfalica (Kolbe)	bark psocid
Cerodontha ireos (Goureau)	(fly) *larva* = iris leaf miner
Cerura vinula (L.)	puss moth
Ceruraphis malifoliae: not *malifoliae* (Fitch, 1855), although misidentified as such in several later published works, see *Dysaphis plantaginea*	
Cetonia aurata (L.)	rose chafer (beetle)
Ceuthorhynchus: misspelling, see *Ceutorhynchus*	

SCIENTIFIC NAMES—Arthropods

Scientific Name   **Common Name**

Ceutorhynchus assimilis (Paykull) .. .. cabbage seed weevil
Ceutorhynchus contractus (Marsham) .. turnip stem weevil
Ceutorhynchus picitarsis Gyllenhal .. .. rape winter stem weevil
   Ceutorhynchus pallidactylus (Marsham), see *Ceutorhynchus quadridens*
Ceutorhynchus pleurostigma (Marsham) .. turnip gall weevil
Ceutorhynchus quadridens (Panzer) .. cabbage stem weevil
Chaetocnema concinna (Marsham) .. .. mangold flea beetle
                                     mangel flea beetle
                                       beet flea beetle
Chaetosiphon fragaefolii (Cockerell) .. strawberry aphid
Chaitophorus leucomelas Koch .. .. poplar leaf aphid
Chaitophorus populeti (Panzer) .. .. poplar shoot aphid
CHALCIDOIDEA .. .. chalcids (wasps)
Chalcoides aurata (Marsham) .. .. willow flea beetle
Chalcoides aurea (Fourcroy) .. .. .. poplar flea beetle
Chamaepsila, subgenus of *Psila*
   Charaeas graminis (L.), see *Cerapteryx graminis*
   Cheimatobia brumata (L.), see *Operophtera brumata*
Cheiridium museorum (Leach) .. .. book pseudoscorpion
Chelifer cancroides (L.) .. .. .. barn pseudoscorpion
   CHELONETHIDA, see PSEUDOSCORPIONES
Chelopistes meleagridis (L.) .. .. .. large turkey louse
   Chelopistes stylifer (Nitzsch), see *Chelopistes meleagridis*
   Chermes, see *Adelges*
Cheyletiella parasitivorax (Mégnin) .. rabbit fur mite
   see also *Listrophorus gibbus* Pagenstecher
Cheyletiella yasguri Smiley .. .. .. dog fur mite
†*Chilocorus* spp. .. .. black ladybirds (beetles)
CHILOPODA .. .. centipedes
   Chion cinctus (Drury), see *Knulliana cinctus*
   Chionaspis alni Signoret, see *Chionaspis salicis*
   Chionaspis aspidistrae Signoret, see *Pinnaspis aspidistrae*
   Chionaspis populi (Baerensprung), see *Chionaspis salicis*
Chionaspis salicis (L.) .. .. .. willow scale
                                         willow scurfy scale
Chirodiscoides caviae Hirst .. .. .. guinea pig fur mite
                                       guinea pig mite
CHIRONOMIDAE .. .. non-biting midges
                                       larvae (when red) = bloodworms
Chirothrips hamatus Trybom .. .. meadow foxtail thrips
Chirothrips manicatus Haliday .. .. grass flower thrips
   Chloridea obsoleta: not *obsoleta* (Fabricius, 1793), although misidentified as such in Hampson, 1903, and in several later published works, see *Helicoverpa armigera*
Chloroclysta truncata (Hufnagel) .. .. common marbled carpet moth
Chloroclystis rectangulata (L.) .. .. green pug moth
   Chlorophorus annularis (Fabricius), see *Rhaphuma annularis*
   Chloropisca circumdata (Meigen), see *Thaumatomyia notata*
Chlorops pumilionis (Bjerkander) .. .. gout fly
Chlorops taeniopus Meigen, see *Chlorops pumilionis*
Chloropulvinaria floccifera (Westwood) .. cushion scale
Cholodkovskya, subgenus of *Adeleges*
Choneiulus palmatus (Nemec) .. .. palm millepede
†*Chorebus gracilis* (Nees) .. .. .. carrot fly parasite (wasp)
Choreutis pariana (Clerck) .. .. .. (moth) *larva* = apple leaf skeletonizer
Chorioptes bovis (Hering) .. .. .. chorioptic mange mite
                                       horse foot mange mite
                                       ox tail mange mite
                                       sheep chorioptic mange mite
                                       symbiotic mange mite

SCIENTIFIC NAMES — Arthropods

Scientific Name	Common Name
Chorioptes caprae (Delafond & Bourguignon)	
Chorioptes communis (Cave)	
Chorioptes cuniculi Railliet	see *Chorioptes bovis*
Chorioptes equi (Gerlach)	
Chorioptes ovis (Zürn)	
Chortophila brassicae (Bouché), see *Delia radicum*	
Chortophila gnava (Meigen), see *Botanophila gnava*	
Chortophila sepia (Meigen), see *Phorbia sepia*	
Chromaphis juglandicola (Kaltenbach)	small walnut aphid
*Chrysodeixis chalcites (Esper)	golden twin spot moth
CHRYSIDIDAE	ruby-tailed wasps
	cuckoo wasps
Chrysolina menthastri (Suffrian)	mint leaf beetle
Chrysomela populi (L.)	red poplar leaf beetle
CHRYSOMELIDAE	leaf beetles
*Chrysomphalus aonidum (L.)	Florida red scale
Chrysomphalus ficus Ashmead, see *Chrysomphalus aonidum*	
†*Chrysopa* spp.	green lacewings
†*Chrysopa carnea* Stephens	common green lacewing
†*Chrysopa perla* (L.)	pearly green lacewing
Chrysops spp.	deer flies
Chrysops caecutiens (L.)	blinding breeze fly
Chrysoteuchia culmella (L.)	garden grass veneer moth
	grass moth
Cicadella aurata (L.), see *Eupteryx aurata*	
Cicadella melissae (Curtis), see *Eupteryx melissae*	
Cicadella stellulata (Burmeister), see *Aguriahana stellulata*	
CICADELLIDAE	leafhoppers
Cicindela campestris L.	green tiger beetle
CICINDELINAE	tiger beetles
Cimex columbarius Jenyns	pigeon bug
	fowl bug
Cimex lectularius L.	bed bug
	wall-louse
Cimex pipistrelli Jenyns	bat bug
Cinara boerneri Hille Ris Lambers, see *Cinara cuneomaculata*	
Cinara cuneomaculata (del Guercio)	larch aphid
Cinara cupressi (Buckton)	cypress aphid
Cinara fresai Blanchard	American juniper aphid
Cinara kochiana (Börner)	giant larch aphid
Cinara pectinate (Nördlinger)	green striped fir aphid
Cinara piceae (Panzer)	spruce stem aphid
	spruce bark aphid
Cinara pilicornis (Hartig)	brown spruce aphid
	spruce shoot aphid
Cinara pinea (Mordvilko)	large pine aphid
Cinara stroyani (Pašek)	green-striped spruce bark aphid
Cinara tujafilina (del Guercio)	thuya aphid
Cionus scrophulariae (L.)	figwort weevil
Cladius difformis (Jurine in Panzer)	antler sawflies
Cladius pectinicornis (Fourcroy)	
Cladius viminalis (Fallén), see *Trichiocampus viminalis*	
Cleonus piger: misspelling see *Cleonus pigra*	
Cleonus pigra (Scopoli)	thistle root gall beetle
Clepsis costana: not *costana* (Denis & Schiffermüller, 1775), although misidentified as such in Fabricius, 1787, and in several later published works, see *Clepsis spectrana*	
Clepsis senecionana (Hübner)	rustic tortrix (moth)
	rustic twist

SCIENTIFIC NAMES—Arthropods

Scientific Name	Common Name
Clepsis spectrana (Treitschke)	straw-coloured tortrix (moth) cyclamen tortrix fern tortrix
Closterotomus, subgenus of *Calocoris*	
Clytus arietis (L.)	wasp beetle
Cnemidocoptes, see *Knemidokoptes*	
Cneorhinus plagiatus (Schaller), see *Philopedon plagiatus*	
Cnephasia asseclana (Denis & Schiffermüller)	flax tortrix (moth) *larva* = poppy leaf roller
Cnephasia incertana (Treitschke)	light grey tortrix (moth) allied shade moth
Cnephasia interjectana (Haworth), see *Cnephasia asseclana*	
Cnephasia longana (Haworth)	(moth) *larva* = omnivorous leaf tier
Cnephasia stephensiana (Doubleday)	grey tortrix (moth)
Cnephasia virgaureana (Treitschke), see *Cnephasia asseclana*	
Cnephasiella, subgenus of *Cnephasia*	
COCCIDAE see also DIASPIDIDAE and MARGARODIDAE	{ scale insects { soft scales
Coccinella 7-punctata L., see *Coccinella septempunctata*	
Coccinella 11-punctata L., see *Coccinella undecimpunctata*	
†*Coccinella quinquepunctata* L.	five-spot ladybird (beetle)
†*Coccinella septempunctata* L.	seven-spot ladybird (beetle)
†*Coccinella undecimpunctata* L.	eleven-spot ladybird (beetle)
COCCINELLIDAE	ladybirds (beetles)
Coccus hesperidum L. see also *Eulecanium tiliae*	brown soft scale soft scale (see also COCCIDAE)
**Cochliomyia hominivorax* (Coquerel)	(fly)*larva* = screw worm New World screw worm
Coelopa frigida (Fabricius) } *Coelopa pilipes* Haliday }	seaweed flies
Coleophora anatipennella (Hübner)	(moth) *larva* = pistol casebearer
Coleophora coracipennella (Hübner), see *Coleophora spinella*	
Coleophora laricella (Hübner)	(moth) *larva* = larch casebearer larch leaf miner
Coleophora nigricella (Stephens), see *Coleophora spinella*	
Coleophora spinella (Schrank)	(moth) *larva* = apple and plum casebearer cigar casebearer
COLEOPHORIDAE	(moth) *larvae* = casebearers
COLEOPTERA	beetles (includes weevils)
COLLEMBOLA	springtails
Colomerus vitis (Pagenstecher)	vine leaf blister mite grape erineum mite vine erineum mite
Coloradoa rufomaculata (Wilson)	small chrysanthemum aphid
Columbicola bacula (Nitzsch), see *Columbicola columbae*	
Columbicola columbae (L.)	pigeon wing louse slender pigeon louse
Columbicola filiformis (von Olfers), see *Columbicola columbae*	
**Comstockaspis perniciosa* (Comstock)	San José scale Chinese scale pernicious scale
Conicera tibialis Schmitz	coffin fly
†*Coniopteryx* spp. see also *Conwentzia* spp. and *Semidalis aleyrodiformis*	powdery lacewings
Conopia myopaeformis (Borkhausen), see *Synanthedon myopaeformis*	
**Conotrachelus nenuphar* (Herbst)	plum curculio (weevil)
Contarinia baeri Prell	pine needle gall midge
Contarinia chrysanthemi (Kieffer)	shasta daisy midge

SCIENTIFIC NAMES—Arthropods

Scientific Name | Common Name

Contarinia dactylidis (Loew) .. .. cocksfoot midge
   see also *Contarinia geniculati, Contarinia merceri, Dasineura dactylidis* and *Sitodiplosis dactylidis*
Contarinia geniculati (Reuter) .. .. { cocksfoot midge / foxtail
   see also *Contarinia dactylidis,*
      Contarinia merceri, Dasineura alopecuri,
      Dasineura dactylidis and *Sitodiplosis dactylidis*
Contarinia humuli (Theobald) .. .. hop strig midge
                                  larva = hop strig maggot
   Contarinia humuli Tölg: attribution to author incorrect, see *Contarinia humuli* (Theobald)
Contarinia loti (Degeer) .. .. .. trefoil flower midge
Contarinia medicaginis Kieffer .. .. lucerne flower midge
Contarinia merceri Barnes .. .. .. { cocksfoot midge / foxtail midge
   see also *Contarinia dactylidis,*
      Contarinia geniculati, Dasineura alopecuri,
      Dasineura dactylidis and *Sitodiplosis dactylidis*
Contarinia nasturtii (Kieffer) .. .. swede midge
Contarinia onobrychidis Kieffer .. .. sainfoin flower midge
Contarinia petioli (Kieffer) .. .. .. poplar gall midge
Contarinia pisi (Winnertz) .. .. .. pea midge
Contarinia pyrivora (Riley) .. .. .. pear midge
Contarinia rubicola Kieffer .. .. .. blackberry flower midge
   Contarinia rubicola Rübsaamen, see *Contarinia rubicola*
Contarinia tiliarum (Kieffer) .. .. .. lime leaf-petiole gall midge
Contarinia tritici (Kirby) .. .. .. yellow wheat blossom midge
                                    wheat blossom midge (see also *Sitodiplosis mosellana*)
†*Conwentzia* spp. .. .. powdery lacewings
   see also *Coniopteryx* spp. and *Semidalis aleyrodiformis*
   Copidosoma tortricis Waterstone, see *Litomastix aretas*
Corcyra cephalonica (Stainton) .. .. rice moth
**Cordylomera spinicornis* (Fabricius) } .. African metallic-green longhorn beetles
**Cordylomera suturalis* Chevrolat }
Corylobium avellanae (Schrank) .. .. large hazel aphid
   Corymbites spp., see *Ctenicera* spp.
Cosmoglyphus laarmani Samsinak .. .. copra itch mite
Cossus cossus (L.) .. .. goat moth
   Cossus ligniperda Fabricius, see *Cossus cossus*
   CRAMBIDAE, see CRAMBINAE
CRAMBINAE .. .. grass moths
   Crambus hortuella(Hübner), see *Chrysoteuchia culmella*
   Crambus hortuellus(Hübner), see *Chrysoteuchia culmella*
Crataerina pallida (Latreille) .. .. .. swift and swallow parasitic fly
   Cratichneumon nigritarius (Gravenhorst), see *Cratichneumon viator*
†*Cratichneumon viator* (Scopoli) .. .. pine looper parasite (wasp)
   see also *Dusona oxyacanthae, Heteropelma calcator* and *Polytribax arrogans*
   Crepidodera aurata (Marsham), see *Chalcoides aurata*
Crepidodera ferruginea (Scopoli) .. .. wheat flea beetle
   Crepidodera helxines: not *helxines* (Linnaeus, 1758), although misidentified as such in several later published works, see *Chalcoides aurea*
   Criocephalus rusticus (L.), see *Arhopalus rusticus*
Crioceris asparagi (L.) .. .. .. asparagus beetle
   Crioceris lilii (Scopoli), see *Lilioceris lilii*
Croesus septentrionalis (L.) .. .. .. hazel sawfly
Croesus varus (de Villaret) .. .. .. alder sawfly
CRUSTACEA .. .. crustaceans (woodlice, shrimps, etc.)
   Cryptococcus fagi (Baerensprung), see *Cryptococcus fagisuga*
Cryptococcus fagisuga Lindinger .. .. beech scale
                                    felted beech scale
   Cryptohypnus riparius (Fabricius), see *Hypnoidus riparius*

SCIENTIFIC NAMES—Arthropods

Scientific Name	Common Name
Cryptolestes ferrugineus (Stephens)	rust-red grain beetle
Cryptolestes minutus (Olivier), not *minutus* (Fourcroy, 1785), see *Cryptolestes pusillus*	
Cryptolestes pusillus (Schoenherr)	flat grain beetle
Cryptolestes turcicus (Grouvelle)	Turkish grain beetle
Cryptomyzus galeopsidis (Kaltenbach)	black currant aphid / currant blister aphid
Cryptomyzus ribis (L.)	red currant blister aphid
CRYPTOPHAGIDAE	mould beetles
*Cryptophlebia leucotreta (Meyrick)	false codling moth
Cryptorhynchus lapathi (L.)	osier weevil / willow beetle / *larva* = willow borer
CRYPTOSTIGMATA	beetle mites / oribatid mites
Ctenarytaina eucalypti (Maskell)	eucalyptus sucker / blue-gum sucker
Ctenicera spp	upland click beetles / *larvae* = upland wireworms
Ctenocephalides canis (Curtis)	dog flea
Ctenocephalides felis (Bouché)	cat flea
Cuclotogaster eynsfordi (Theobald), see *Cuclotogaster heterographus*	
Cuclotogaster heterographus (Nitzsch)	chicken head louse / neck louse
Culex pipiens L.	common gnat / house gnat
CULICIDAE	gnats and mosquitoes
Culicoides spp.	biting midges
Culiseta annulata (Schrank)	banded mosquito
Curculio glandium Marsham } *Curculio venosus* (Gravenhorst) }	acorn weevils
Curculio nucum L.	hazel nut weevil / nut weevil
†*Curculio salicivorus* Paykull	willow bean-gall sawfly predator (weevil)
CURCULIONIDAE	weevils
Cydia conicolana (Heylaerts)	pine cone moth
Cydia coniferana (Ratzeburg)	pine resin moth
Cydia fagiglandana (Zeller)	beech seed moth
Cydia funebrana (Treitschke)	plum fruit moth / *larva* = red plum maggot
*Cydia molesta (Busck)	oriental fruit moth
Cydia nigricana (Fabricius)	pea moth
Cydia pomonella (L.)	codling moth / *larva* = apple worm
Cydia splendana (Hübner)	acorn moth
CYDNIDAE	shield bugs
see also ACANTHOSOMIDAE, PENTATOMIDAE and SCUTELLERIDAE	
Cylindroiulus britannicus (Verhoeff)	lesser glasshouse millepede
Cylindroiulus londinensis (Leach)	black millepede
see also *Tachypodoiulus niger*	snake millepede (see also *Archiboreoiulus pallidus*, *Boreoiulus tenuis* and *Blaniulus guttulatus*)
Cylindroiulus punctatus (Leach)	woodland floor millepede
Cylindroiulus teutonicus (Pocock), see *Cylindroiulus londinensis*	
CYNIPOIDEA	cynipids / gall wasps
Cynips divisa Hartig	oak bud red-gall cynipid (wasp)
Cynips kollari (Hartig), see *Andricus kollari*	
Cynips longiventris Hartig	oak leaf striped-gall cynipid (wasp)
Cynips quercusfolii L.	oak leaf cherry-gall cynipid (wasp)
Cytodites nudus (Vizioli)	air-sac mite

SCIENTIFIC NAMES—Arthropods

Scientific Name **Common Name**

†*Dacnusa* spp. .. .. chrysanthemum leaf miner parasites (wasps)
  see also *Diglyphus* spp.
  Dacnusa gracilis (Nees), see *Chorebus gracilis*

  Dactylosphaera vitifoliae (Fitch) ⎫
  Dactylosphaera vitifolii (Fitch) ⎬ see *Daktulosphaira vitifoliae*
**Daktulosphaira vitifoliae* (Fitch) .. .. grape phylloxera
                                                  vine louse
  Damalinia bovis (L.), see *Bovicola bovis*
  Damalinia caprae (Gurlt), see *Bovicola caprae*
  Damalinia climax (Nitzsch in Giebel), see *Bovicola caprae*
  Damalinia equi (Denny), see *Werneckiella equi*
  Damalinia ovis (Schrank), see *Bovicola ovis*
  Damalinia ovisarietis (Schrank), see *Bovicola ovis*

  Damalinia parumpilosa (Piaget) ⎫
  Damalinia pilosa (Giebel) ⎬ see *Werneckiella equi*
  Damalinia scalaris (Nitzsch), see *Bovicola bovis*
  Damalinia solida (Rudow), see *Bovicola caprae*
  Damalinia sphaerocephala (von Olfers), see *Bovicola ovis*
  Damalinia tauri (L.), see *Bovicola bovis*
Dasineura affinis (Kieffer) .. .. .. violet leaf midge
Dasineura alni (Löw) .. .. alder leaf gall midge
Dasineura alopecuri (Reuter) .. .. foxtail midge
  see also *Contarinia geniculati* and *Contarinia merceri*
Dasineura alpestris (Kieffer) .. .. .. arabis midge
Dasineura brassicae (Winnertz) .. .. brassica pod midge
                                            bladder pod midge
                                            pod midge
  Dasineura clausilia (Bremi), see *Rhabdophaga clausilia*
Dasineura crataegi (Winnertz) .. .. hawthorn button-top midge
Dasineura dactylidis Metcalfe .. .. cocksfoot midge
  see also species of *Contarinia* and *Sitodiplosis dactylidis*
Dasineura fraxini (Bremi) .. .. .. ash midrib pouch-gall midge
  Dasineura fraxini (Kieffer), see above
Dasineura gentneri Pritchard .. .. white clover seed midge
Dasineura glechomae (Kieffer) .. .. lighthouse-gall midge
Dasineura gleditchiae (Osten-Sacken) .. honeylocust gall midge
Dasineura leguminicola (Lintner) .. .. clover seed midge
                                               clover flower midge
Dasineura mali (Kieffer) .. .. .. apple leaf midge
                                               apple leaf-curling midge
Dasineura plicatrix (Loew) .. .. .. blackberry leaf midge
Dasineura pyri (Bouché) .. .. .. pear leaf midge
                                               pear leaf-curling midge
Dasineura ribicola (Kieffer) .. .. .. gooseberry leaf midge
Dasineura rosarum (Hardy) .. .. .. rose leaf midge
Dasineura tetensi (Rübsaamen) .. .. black currant leaf midge
                                               black currant leaf-curling midge

Dasineura thomasiana (Kieffer) ⎫
Dasineura tiliamvolvens (Rübsaamen) ⎬ .. lime leaf-margin gall midges
Dasineura trifolii (Loew) .. .. .. clover leaf midge
Dasineura urticae (Perris) .. .. .. nettle gall midge
Dasineura viciae (Kieffer) .. .. .. vetch leaf midge
  Dasychira pudibunda (L.), see *Calliteara pudibunda*
  Dasyneura spp., see *Dasineura* spp.
  Dasyphora spp., see *Eudasyphora* spp.
  Dasyphora cyanella (Meigen), see *Eudasyphora cyanella*
Delia antiqua (Meigen) .. .. .. onion fly

SCIENTIFIC NAMES—Arthropods

Scientific Name   **Common Name**

Delia brassicae (Bouché)   ⎱ see *Delia radicum*
Delia brassicae (Wiedemann)   ⎰
Delia brunnescens: not *brunnescens* (Zetterstedt, 1845), although misidentified as such in several later published works, see *Delia cardui*
Delia cardui (Meigen) .. .. .. carnation fly
Delia cilicrura (Rondani), see *Delia platura*
Delia coarctata (Fallén) .. .. .. wheat bulb fly
Delia echinata (Séguy) .. .. .. spinach stem fly
Delia floralis (Fallén) .. .. turnip root fly
Delia florilega (Zetterstedt) ⎱ .. .. .. bean seed flies
Delia platura (Meigen)   ⎰
Delia radicum (L.) .. .. cabbage root fly
                                   larva = cabbage maggot
Delphacodes pellucida (Fabricius), see *Javesella pellucida*
Delphiniobium aconiti (van der Goot), see *Delphiniobium junackianum*
Delphiniobium junackianum (Karsch)   monkshood aphid
Demodex bovis Stiles .. .. cattle follicle mite
                                   cattle mange mite
Demodex canis Leydig .. .. .. dog follicle mite
                                   dog red-mange mite
Demodex caprae Railliet .. .. .. goat follicle mite
                                   goat mange mite
Demodex cati Mégnin .. .. .. cat follicle mite
                                   cat mange mite
Demodex cuniculi (Pfeiffer) .. .. .. rabbit follicle mite
                                   rabbit mange mite
Demodex equi Railliet .. .. .. horse follicle mite
                                   horse mange mite
Demodex folliculorum (Simon) .. .. human follicle mite
Demodex ovis Railliet .. .. .. sheep follicle mite
                                   sheep mange mite
Demodex phylloides Csokor .. .. .. pig follicle mite
                                   pig head mange mite
DEMODICIDAE .. .. follicle mites
   (in particular *Demodex* spp.)
   see also *Psorergates* spp.
**Dendroctonus brevicomis* LeConte .. .. western pine beetle
**Dendroctonus micans* (Kugelann) .. .. great spruce bark beetle
                                   European spruce beetle
Dendromyza cerasiferae Kangas, see *Phytobia cerasiferae*
Dendrothrips ornatus (Jablonowski) .. privet thrips
†*Dendroxena quadrimaculata* (Scopoli) .. four-spotted burying beetle
   Dentatus tulipae (Boyer de Fonscolombe), see *Dysaphis tulipae*
Deporaus betulae (L.) .. .. .. birch leaf roller weevil
Depressaria conterminella Zeller, see *Agonopterix conterminella*
Depressaria heracliana: not *heracliana* (Linnaeus, 1758), although misidentified as such in several later published works, see *Depressaria pastinacella*
Depressaria nervosa Haworth, see *Agonopterix nervosa*
Depressaria pastinacella (Duponchel) .. parsnip moth
                                   cow-parsnip flat-body
                                   larva = parsnip webworm
Dermacentor reticulatus (Fabricius) .. marsh tick
Dermanyssus gallinae (Degeer) .. .. chicken mite
                                   poultry red mite
DERMAPTERA .. .. earwigs
Dermatophagoides spp. .. .. .. house dust mites
Dermatophagoides farinae Hughes .. .. American house dust mite
   Dermatophagoides longior (Trouessart), see *Euroglyphus longior*

SCIENTIFIC NAMES—Arthropods

Scientific Name | Common Name

Dermatophagoides pteronyssinus
 (Trouessart) .. .. European house dust mite
Dermestes spp. .. .. hide beetles
Dermestes haemorrhoidalis Küster .. .. black larder beetle
Dermestes lardarius L. .. .. .. bacon beetle
  larder beetle
Dermestes maculatus Degeer .. .. leather beetle
Dermestes peruvianus Laporte de Castelnau Peruvian larder beetle
 Dermestes vulpinus Fabricius, see *Dermestes maculatus*
 Deudorix antalus (Hopffer), see *Hypokopelates antalus*
Diachrysia orichalcea (Fabricius) .. .. slender burnished brass moth
†*Diadegma fenestralis* (Holmgren) .. .. diamond-back moth parasite (wasp)
†*Diaeretiella rapae* (M'Intosh) .. .. cabbage aphid parasite (wasp)
 Diaeretus rapae (Curtis), see *Diaeretiella rapae*
Dialeurodes chittendeni Laing .. .. rhododendron whitefly
 Diarthronomyia chrysanthemi Ahlberg, see *Rhopalomyia chrysanthemi*
DIASPIDIDAE .. .. { armoured scales
 see also COCCIDAE and   { scale insects
 MARGARODIDAE
Diaspis boisduvalii Signoret .. .. .. orchid scale
**Diaspis bromeliae* (Kerner) .. .. .. pineapple scale
 Diaspis caruelii: description of *caruelii* Targioni-Tozzetti, 1869, is unresolved; in later
  published works the name has been applied to two species: (1) *Carulaspis minima*,
  questionably *caruelii* Targioni-Tozzetti and (2) *Carulaspis juniperi*, a misidentification of
  caruelii Targioni-Tozzetti, see *Carulaspis minima* and *Carulaspis juniperi*
 Diaspis rosae Bouché, see *Aulacaspis rosae*
Diastrophus rubi (Bouché) .. .. .. rose stem gall cynipid (wasp)
 see also *Diplolepis mayri*
 Diataraxia oleracea (L.), see *Lacanobia oleracea*
Dichomeris marginella (Fabricius) .. .. juniper webber moth
   larva = juniper webworm
 Dicranura vinula (L.), see *Cerura vinula*
Dictyonota strichnocera Fieber .. .. gorse and broom lace bug
DICTYOPTERA .. .. cockroaches, etc.
Dicyphus errans (Wolff) .. .. .. slender grey capsid (bug)
 Didymomyia reamuriana (Loew), see *Didymomyia tiliacea*
Didymomyia tiliacea (Bremi) .. .. lime leaf gall midge
Dienerella filum (Aubé) .. .. .. herbarium beetle
†*Diglyphus* spp. .. .. chrysanthemum leaf miner parasites (wasps)
 see also *Dacnusa* spp.
 Dikraneura mollicula (Boheman), see *Emelyanoviana mollicula*
Diloba caeruleocephala (L.) .. .. figure of eight moth
Dilophus febrilis (L.) .. .. .. fever fly
   blossom fly
**Dinoderus bifoveolatus* (Wollaston) .. West African ghoon beetle
*Dinoderus brevis Horn .. .. .. Indian ghoon beetle
**Dinoderus japonicus* Lesne .. .. Japanese ghoon beetle
Dioryctria abietella (Denis and
 Schiffermüller) .. .. pine knothorn moth
 Diplolepis dispar (Niblet), see *Diplolepis nervosa*
Diplolepis eglanteriae (Hartig) .. .. rose smooth pea-gall cynipid (wasp)
Diplolepis mayri (Schlechtendal) .. .. rose stem gall cynipid (wasp)
 see also *Diastrophus rubi*
Diplolepis nervosa (Curtis) .. .. rose spiked pea-gall cynipid (wasp)
Diplolepis rosae (L.) .. .. .. bedeguar gall wasp
   larva induces gall named 'robin's pincushion'
Diplolepis spinosissimae (Giraud) .. .. rose spherical-gall cynipid (wasp)
DIPLOPODA .. .. millepedes
   millipedes

SCIENTIFIC NAMES—Arthropods

Scientific Name   **Common Name**

Diplosis tremulae (Winnertz), see *Harmandia tremulae*
Diprion pini (L.) .. ..   pine sawfly
                                    large pine sawfly
Diprion sertifer (Fourcroy), see *Neodiprion sertifer*
Diptacus gigantorhynchus (Nalepa) .. ..   big-beaked plum mite
                                    plum gall mite
DIPTERA .. ..   flies (includes gnats, midges and mosquitoes)
Discestra trifolii (Hufnagel) .. .. ..   nutmeg moth
Ditula angustiorana (Haworth) .. ..   fruit tree tortrix (moth)
   see also species of *Acleris*, *Apotomis*,   red-barred tortrix
     Archips, *Hedya*, *Olethreutes* and   vine tortrix
     Pandemis
Dixippus morosus (Brunner): attribution to author incorrect, see *Carausius morosus* (Sinéty)
Dizygomyza, subgenus of *Cerodontha*
   Dizygomyza barnesi Hendel, see *Phytobia cambii*
   Dizygomyza carbonaria (Zetterstedt), see *Phytobia cambii*
   Dizygomyza iraeos (Robineau Desvoidy), see *Cerodontha ireos*
   Dizygomyza ireos (Goureau), see *Cerodontha ireos*
†*Dolichonabis limbatus* (Dahlbom) .. ..   marsh damsel bug
Dolichovespula norwegica (Fabricius) ..   Norwegian wasp
Dolichovespula sylvestris (Scopoli) .. ..   tree wasp
Dolycoris baccarum (L.) .. .. ..   sloe bug
   Domomyza spp., see *Agromyza* spp.
Dorcus parallelipipedus (L.) .. .. ..   lesser stag beetle
Drepanosiphum platanoidis (Schrank) ..   sycamore aphid
Dreyfusia, subgenus of *Adelges*
   Dreyfusia nordmannianae (Eckstein), see *Adelges nordmannianae*
   Dreyfusia piceae (Ratzeburg), see *Adelges piceae*
Drosophila spp. .. ..   small fruit flies
                                    vinegar flies
Drosophila funebris (Fabricius) .. ..   milkbottle fly
Drosophila repleta Wollaston .. ..   kitchen fly
DROSOPHILIDAE .. ..   small fruit flies
   (in particular *Drosophila* spp.) .. ..   vinegar flies
Dryocoetes autographus (Ratzeburg) ..   silver fir bark beetle
   Dryocoetes villosus (Fabricius), see *Dryocoetinus villosus*
Dryocoetinus villosus (Fabricius) .. ..   (beetle) *larva* = brown punctured borer
†*Dusona oxyacanthae* (Boie) .. .. ..   pine looper parasite (wasp)
   see also *Cratichneumon viator*, *Heteropelma calcator* and *Polytribax arrogans*
Dynaspidiotus britannicus (Newstead) ..   bay-tree scale
                                    holly scale
Dysaphis apiifolia (Theobald)
   ssp. *petroselini* (Börner) .. ..   hawthorn—parsley aphid
Dysaphis aucupariae (Buckton) .. ..   wild service aphid
Dysaphis crataegi (Kaltenbach) .. ..   hawthorn—carrot aphid
Dysaphis devecta (Walker) .. .. ..   rosy leaf-curling aphid
Dysaphis plantaginea (Passerini) .. ..   rosy apple aphid
                                    blue bug (see also *Zicrona caerulea*)
Dysaphis pyri (Boyer de Fonscolombe) ..   pear—bedstraw aphid
Dysaphis radicola (Mordvilko) .. ..   apple—dock aphid
Dysaphis sorbi (Kaltenbach) .. .. ..   mountain ash aphid
Dysaphis tulipae (Boyer de Fonscolombe)   tulip bulb aphid
Dytiscus spp. .. ..   diving water beetles

Earias clorana (L.) .. ..   osier green moth
                                    cream-bordered green pea moth
**Eburia quadrigeminata* (Say) .. ..   ivory-marked longhorn beetle
   Eccoptogaster: invalid name, see *Scolytus*

SCIENTIFIC NAMES—Arthropods

Scientific Name   **Common Name**

Ectomyelois ceratoniae (Zeller) .. .. locust bean moth
Edwardsiana avellanae (Edwards) ⎫
Edwardsiana crataegi (Douglas)   ⎬ fruit tree leafhoppers
Edwardsiana hippocastani (Edwards) ⎬
Edwardsiana prunicola (Edwards) ⎭
   see also species of *Alnetoidia, Ribautiana, Typhlocyba* and *Zygina*
   Edwardsiana flavescens (Fabricius), see *Empoasca vitis*
Edwardsiana rosae (L.) .. .. .. rose leafhopper
*Elaphidion mucronatum (Say) ⎫ .. .. American twig-pruner beetles
*Elaphidion nanum (Fabricius) ⎭
Elasmostethus interstinctus (L.) .. .. birch bug
Elasmucha grisea (L.) .. .. .. parent bug
                                      mothering bug
ELATERIDAE .. .. click beetles
   (in particular species of *Agriotes* and   skipjacks
     Athous spp.)   *larvae* = wireworms
Elatobium abietinum (Walker) .. .. green spruce aphid
                                        spruce aphid
   Ellopia fasciaria (L.), see *Hylaea fasciaria*
Elophila nymphaeata (L.) .. .. .. brown china mark moth
Emelyanoviana mollicula (Boheman) .. thyme leafhopper
   Emphytus carpini (Hartig), see *Protemphytus carpini*
   Emphytus cinctus (L.), see *Allantus cinctus*
   Emphytus pallipes (Spinola), see *Protemphytus pallipes*
   Empleurus nubilus (Fabricius), see *Helophorus nubilus*
Empoasca decipiens Paoli ⎫
Empoasca vitis (Göthe)   ⎬.. .. .. green leafhoppers
                                        green frogflies
   Empoasca flavescens (Fabricius), see *Empoasca vitis*
Empria tridens (Konow) .. .. .. raspberry sawfly
*Enaphalodes rufulum (Haldeman) .. .. (beetle) *larva* = red oak borer
Enarmonia formosana (Scopoli) .. .. cherry bark tortrix (moth)
   Enarmonia woeberiana (Denis & Schiffermüller), see *Enarmonia formosana*
†*Encarsia formosa* Gahan .. .. .. glasshouse whitefly parasite (wasp)
Endelomyia aethiops (Fabricius) .. .. rose slug sawfly
                                        rose sawfly
   Endrosis lactella (Denis & Schiffermüller), see *Endrosis sarcitrella*
Endrosis sarcitrella (L.) .. .. .. white-shouldered house moth
   Entodecta, see *Metallus*
Eoseristalis, subgenus of *Eristalis*
Eotetranychus tiliarium (Hermann) .. .. lime mite
                                        linden mite
EPHEMEROPTERA .. .. .. mayflies
Ephestia calidella Guenée .. .. .. carob moth
                                        dried fruit moth
Ephestia cautella (Walker) .. .. .. tropical warehouse moth
                                        almond moth
                                        dried currant moth
   Ephestia defectella (Walker), see *Ephestia cautella*
Ephestia elutella (Hübner) .. .. .. warehouse moth
                                        cacao moth
                                        stored tobacco moth
Ephestia figulilella Gregson .. .. .. raisin moth
                                        fig moth
Ephestia kuehniella Zeller .. .. .. Mediterranean flour moth
                                        mill moth
   Ephestia sericarium (Scott), see *Ephestia elutella*
   Ephialtes pomorum (Ratzeburg), see *Scambus pomorum*
EPHYDRIDAE .. .. shore flies

SCIENTIFIC NAMES—Arthropods

Scientific Name	Common Name
Epiblema uddmanniana (L.)	bramble shoot moth
	larva = bramble shoot webber
**Epichoristodes acerbella* (Walker)	African carnation tortrix (moth)
Epichoristodes iocoma (Meyrick), see *Epichoristodes acerbella*	
Epichoristodes ionephela (Meyrick), see *Epichoristodes acerbella*	
EPIDERMOPTIDAE	bird ked mites
	scab mites (see also PSOROPTIDAE and PYROGLYPHIDAE)
Epinotia cruciana (L.)	willow tortrix (moth)
Epinotia nanana (Treitschke)	pine tortrix (moth)
Epinotia tedella (Clerck)	spruce needle tortrix (moth)
	larva = spruce needle miner
Epinotia tenerana (Denis and Schiffermüller)	nut bud tortrix (moth)
Epiphyas postvittana (Walker)	light brown apple moth
	larva = apple leaf roller
Epipolaeus caliginosus (Fabricius), see *Plinthus caliginosus*	
Epirrita autumnata (Borkhausen)	autumn moth
Epirrita dilutata (Denis & Schiffermüller)	November moth
Episema caeruleocephala (L.) see *Diloba caeruleocephala*	
Epitrimerus alinae Liro	chrysanthemum leaf rust mite
	chrysanthemum leaf mite
	chrysanthemum russet mite
	chrysanthemum stem mite
Epitrimerus gigantorhynchus (Nalepa), see *Diptacus gigantorhynchus*	
Epitrimerus piri (Nalepa)	pear rust mite
Epitrimerus pyri: misspelling see *Epitrimerus piri*	
Erannis aurantiaria (Hübner), see *Agriopis aurantiaria*	
Erannis defoliaria (Clerk)	mottled umber moth
*Ergates faber (L.)	Syrian conifer longhorn beetle
**Ergates spiculatus* (LeConte)	American conifer longhorn beetle
Eriococcus devoniensis (Green)	heather scale
Eriogaster lanestris (L.)	small eggar moth
Eriogaster populi (L.), see *Poecilocampa populi*	
Erioischia brassicae (Bouché) } see *Delia radicum*	
Erioischia brassicae (Wiedemann)	
Erioischia floralis (Fallén), see *Delia floralis*	
☆*Eriophyes avellanae* (Nalepa), see *Phytoptus avellanae*	
Eriophyes gallarumtiliae (Turpin), see *Eriophyes tiliae*	
Eriophyes gracilis (Nalepa), see *Phyllocoptes gracilis*	
Eriophyes laevis (Nalepa)	alder bead-gall mite
Eriophyes macrochelus (Nalepa), see *Aceria macrochelus*	
☆*Eriophyes padi* (Nalepa)	plum leaf gall mite
Eriophyes psilaspis (Nalepa), see *Cecidophyopsis psilaspis*	
Eriophyes pyri (Pagenstecher)	pear leaf blister mite
Eriophyes ribis (Westwood), see *Cecidophyopsis ribis*	
Eriophyes rudis (Canestrini), see *Acalitus rudis*	
Eriophyes similis (Nalepa)	plum pouch-gall mite
Eriophyes tiliae (Pagenstecher)	nail gall mite
	tin-tack gall mite
Eriophyes triradiatus (Nalepa)	witches' broom mite (of willow)
Eriophyes tristriata (Nalepa), see *Aceria tristriatus*	
Eriophyes vitis (Pagenstecher), see *Colomerus vitis*	

☆Note: the use of the generic names *Phytoptus* and *Eriophyes* follows that in general use before 1971, and not that proposed by Newkirk, R. A. and Keifer, H. H., (1971) *Eriophyid Studies* C5, California Department of Agriculture, Sacramento.

SCIENTIFIC NAMES—Arthropods

Scientific Name	Common Name
ERIOPHYIDAE } ERIOPHYOIDEA }	gall mites bud and rust mites eriophyid mites
Eriosoma lanigerum (Hausmann)	woolly aphid American blight aphid
Eriosoma lanuginosum (Hartig)	elm balloon-gall aphid
Eriosoma ulmi (L.)	currant root aphid elm—currant aphid elm leaf aphid
Eristalis pertinax (Scopoli) see also *Eristalis tenax*	(fly) *larva* = rat-tailed maggot
Eristalis tenax (L.) see also *Eristalis pertinax*	drone fly *larva* = rat-tailed maggot

Ernarmonia formosana (Scopoli), see *Enarmonia formosana*
Ernarmonia funebrana (Treitschke), see *Cydia funebrana*
Ernarmonia pomonella (L.), see *Cydia pomonella*

Ernobius mollis (L.) .. .. false furniture beetle
Erythrapion, subgenus of *Apion*
　Erythroneura alneti (Dahlbom), see *Alnetoidia alneti*
　Erythroneura coryli (Tollin): *coryli* (Tollin) is a species synonym, but the name is retained for a subspecies, see *Alnetoidia alneti*
　Erythroneura flammigera (Geoffroy in Fourcroy), see *Zygina flammigera*
　Erythroneura pallidifrons (Edwards), see *Hauptidia maroccana*
　Euacanthus interruptus (L.), see *Evacanthus interruptus*

Eucallipterus tiliae (L.)	lime leaf aphid
Euceraphis betulae (Koch)	silver birch aphid
Euceraphis punctipennis (Zetterstedt)	downy birch aphid birch aphid

Eucosma cruciana (L.), see *Epinotia cruciana*
Eucosma diniana (Guenée), see *Zeiraphera diniana*
Eucosma griseana: not *griseana* (Hübner, 1796-99), although misidentified as such in Fletcher, 1939, and in several later published works, see *Zeiraphera diniana*
Eucosma penkleriana: not *penkleriana* (Denis & Schiffermüller, 1775), although misidentified as such in several later published works, see *Epinotia tenerana*
Eucosma tedella (Clerck), see *Epinotia tedella*

Eudasyphora spp. see also *Hydrotaea* spp., 　*Ophyra* spp., *Pollenia* spp., 　and *Thaumatomyia* spp.	cluster flies swarming flies
Eudasyphora cyanella (Meigen)	green cluster flies
Eulachnus agilis (Kaltenbach)	spotted pine aphid
Eulachnus brevipilosus Börner	narrow green pine aphid
Eulachnus rileyi (Williams)	narrow brown pine aphid
Eulecanium ciliatum (Douglas) see also *Kermes quercus*	oak scale

Eulecanium corni (Bouché), see *Parthenolecanium corni*
Eulecanium coryli (L.), see *Eulecanium tiliae*
Eulecanium crudum Green, see *Parthenolecanium corni*
Eulecanium pomeranicum (Kawecki), see *Parthenolecanium pomeranicum*

Eulecanium tiliae (L.)	nut scale brown soft scale (see also *Coccus hesperidum*)
Euleia heraclei (L.)	celery fly
Eulithis prunata (L.)	phoenix (moth)
†*Eulophus pennicornis* Nees } †*Eulophus larvarum* (L.) }	(wasp) pupae = tombstone chalcids
Eumerus strigatus (Fallén) } *Eumerus tuberculatus* (Rondani) }	small narcissus flies lesser bulb flies

SCIENTIFIC NAMES—Arthropods

Scientific Name   **Common Name**

Eumusca, subgenus of *Musca*
Euophryum confine (Broun)   New Zealand wood-boring weevils
*Euophryum rufum (Broun)   wood-boring weevils (see also
                                     Pentarthrum huttoni)
**Euphranta japonica* Ito .. .. .. Japanese cherry fruit fly
    Eupista anatipennella (Hübner), see *Coleophora anatipennella*
Eupithecia assimilata Doubleday .. .. currant pug moth
Eupithecia nanata (Hübner) .. .. .. narrow-winged pug moth
Eupoecilia ambiguella (Hübner) .. .. vine moth
Euproctis chrysorrhoea (L.) .. .. .. brown-tail moth
    note: the name *chrysorrhoea* was wrongly used in Haworth, 1803, and in several later published works, to describe *Euproctis similis* (Fuessly)
    Euproctis phaeorrhoeus (Haworth), see *Euproctis chrysorrhoea*
Euproctis similis (Fuessly) .. .. .. yellow-tail moth
                                         gold-tail moth
    Eupteroidea stellulata (Burmeister), see *Aguriahana stellulata*
Eupterycyba jucunda (Herrich-Schäffer)
Eupteryx aurata (L.)   .. potato leafhoppers
Eupteryx melissae Curtis .. .. .. chrysanthemum leafhopper
    Eupteryx stellulata (Burmeister), see *Aguriahana stellulata*
    Eurhodope suavella (Zincken), see *Numonia suavella*
    Euribia zoe (Meigen), see *Trypeta zoe*
Euroglyphus maynei (Cooreman) .. .. mattress dust mite
Euroglyphus longior (Trouessart) .. .. granary mite
    Eurytoma gibba Boheman, see *Bruchophagus gibbus*
Eurytoma orchidearum (Westwood) .. orchid fly (wasp)
                                         cattleya 'fly'
    Euschesis comes (Hübner), see *Noctua comes*
Euscrobipalpa, subgenus of *Scrobipalpa*
    Eutromula pariana (Clerck), see *Choreutis pariana*
Euura mucronata (Hartig) .. .. .. willow bud sawfly
    Euura saliceti: not *saliceti* (Fallén, 1808), although misidentified as such in several later published works, see *Euura mucronata*
Euxoa spp. .. .. dart moths
    see also *Agrotis* spp.
                          larvae = cutworms
                                 surface   see also
                                 caterpillars   *Noctua*
                                                   pronuba
Euxoa nigricans (L.) .. .. garden dart moth
Euxoa tritici (L.) .. .. white-line dart moth
    Euzetes lapidarius Michael, see *Humerobates rostrolamellatus*
Evacanthus interruptus (L.) .. .. .. hop leafhopper
                                       hop froghopper
Evergestis forficalis (L.) .. .. .. garden pebble moth
Evetria, subgenus of *Epinotia*
    Evetria buoliana (Denis & Schiffermüller), see *Rhyacionia buoliana*
    Evetria resinella (L.), see *Petrova resinella*
    Evetria turionana (Haworth), see *Blastesthia turionella*
†*Exochomus quadripustulatus* (L.) .. .. conifer ladybird (beetle)
    see also *Aphidecta obliterata*

Fagocyba cruenta (Herrich-Schäffer) .. beech leafhopper
Fannia canicularis (L.) .. .. .. lesser house fly
Fannia scalaris (Fabricius) .. .. .. latrine fly
    Felicola subrostrata (Nitzsch): attribution to author incorrect, see *Felicola subrostratus* (Burmeister)
Felicola subrostratus (Burmeister) .. .. cat biting louse

SCIENTIFIC NAMES—Arthropods

Scientific Name	Common Name

Fenusa albipes Cameron, see *Metallus albipes*
Feronia cuprea (L.), see *Pterostichus cupreus*
Feronia madida (Fabricius), see *Pterostichus madidus*
Feronia melanarius (Illiger), see *Pterostichus melanarius*
Forficula auricularia L. .. .. .. common earwig
Formica fusca L. ⎫
Formica lemani Bondroit ⎭ .. .. .. large black ants
Formica lugubris Zetterstedt .. .. hairy wood ant
†*Formica rufa* L. .. .. wood ant
FORMICIDAE .. .. ants
Frankliniella intonsa (Trybom) .. .. flower thrips
Frankliniella iridis (Watson) .. .. .. iris thrips
**Frankliniella occidentalis* (Pergande) .. western flower thrips
   Frankliniella robusta (Uzel), see *Kakothrips pisivorus*
**Frankliniella schultzei* (Trybom) .. .. cotton bud thrips

Galerucella lineola (Fabricius) .. .. brown willow leaf beetle
Galerucella nymphaeae (L.) .. .. .. water-lily beetle
   Galerucella viburni (Paykull), see *Pyrrhalta viburni*
Galleria mellonella (L.) .. .. .. honeycomb moth
                                         greater wax moth
                                         wax moth
Gasterophilus haemorrhoidalis (L.) .. .. nose bot fly
Gasterophilus intestinalis (Degeer) .. horse bot fly
Gasterophilus nasalis (L.) .. .. .. throat bot fly
Gastrodes abietum Bergroth .. .. .. spruce bug
Gastropacha quercifolia (L.) .. .. .. lappet moth
Gastrophysa polygoni (L.) .. .. .. polygonum leaf beetle
   Geoktapia pyraria (Passerini), see *Melanaphis pyraria*
GEOMETRIDAE .. .. geometer moths
                                         larvae = looper caterpillars
Geomyza tripunctata Fallén .. .. .. grass and cereal fly
   see also *Opomyza* spp. .. .. ..
GEOPHILIDAE .. .. wire centipedes
                                         garden centipedes (see also
                                              LITHOBIIDAE
                                         in particular *Lithobius* spp.)
Geotrupes spp. .. .. dor beetles
Gephyraulus raphanistri (Kieffer) .. .. brassica flower midge
Gibbium psylloides (de Czenpinski) .. shiny spider beetle
Gilletteella, subgenus of *Adelges*
   Gilletteella cooleyi (Gillette), see *Adelges cooleyi*
Gilpinia hercyniae (Hartig) .. .. .. European spruce sawfly
Glomeris marginata (Villers) .. .. pill millepede
   Glycyphagus destructor (Schrank), see *Lepidoglyphus destructor*
Glycyphagus domesticus (Degeer) .. .. house mite
                                         furniture mite
                                         grocers' itch mite (see also *Lepidoglyphus*
                                              destructor)
   Glyphipterix cramerella: not *cramerella* (Fabricius, 1777), although misidentified as such in
      several later published works, see *Glyphipterix simpliciella*
   Glyphipterix fischeriella (Zeller), see *Glyphipterix simpliciella*
Glyphipterix simpliciella (Stephens) .. .. cocksfoot moth
   Glyphipteryx simpliciella (Stephens): misspelling, see entry above
   Gnathocerus: misspelling, see *Gnatocerus*
Gnatocerus cornutus (Fabricius) .. .. broad-horned flour beetle
Gnatocerus maxillosus (Fabricius) .. .. slender-horned flour beetle

SCIENTIFIC NAMES—Arthropods

Scientific Name	Common Name
Gnorimoschema operculella (Zeller), see *Phthorimaea operculella*	
Gohieria fusca (Oudemans)	brown flour mite
Goniocotes gallinae (Degeer)	chicken fluff louse
Goniocotes hologaster (Nitzsch), see *Goniocotes gallinae*	
Goniodes abdominalis (Piaget), see *Goniodes gigas*	
Goniodes dissimilis Denny	brown chicken louse
Goniodes gigas (Taschenberg)	large chicken louse
Goniodes hologaster: not *hologaster* (Nitzsch, 1818), although misidentified as such in Denny, 1842, and in several later published works, see *Goniodes gigas*	
Gonodontis bidentata (Clerck), see *Odontopera bidentata*	
Gortyna flavago (Denis & Schiffermüller)	frosted orange moth
Gortyna micacea (Esper), see *Hydraecia micacea*	
Gracilia minuta (Fabricius)	wicker longhorn beetle
see also *Nathrius brevipennis*	
Gracillaria, subgenus of *Caloptilia*	
Gracillaria azaleella Brants, see *Caloptilia azaleella*	
Gracillaria syringella (Fabricius), see *Caloptilia syringella*	
Graphocephala coccinea (Förster), see *Graphocephala fennahi*	
Graphocephala fennahi Young	rhododendron hopper
Grapholita, subgenus of *Cydia*	
Grapholita funebrana (Treitschke), see *Cydia funebrana*	
Grapholita molesta (Busck), see *Cydia molesta*	
GRYLLIDAE	crickets
Gryllotalpa gryllotalpa (L.)	mole cricket
Gryllulus domesticus (L.), see *Acheta domesticus*	
Gryllus campestris L.	field cricket
Gymnetron spp.	gall weevils
Gymnoglyphus, subgenus of *Euroglyphus*	
Gypsonoma aceriana (Duponchel)	poplar cloaked bell moth

Hadena oleracea (L.), see *Lacanobia oleracea*	
Haemaphysalis punctata Canestrini & Fanzago	coastal red tick
Haematobia irritans (L.)	horn fly
Haematobia stimulans (Meigen), see *Haematobosca stimulans*	
Haematobosca stimulans (Meigen)	cattle biting fly
Haematopinus asini (L.)	horse sucking louse
Haematopinus eurysternus (Nitzsch)	short-nosed cattle louse
Haematopinus macrocephalus (Burmeister), see *Haematopinus asini*	
Haematopinus suis (L.)	pig louse
	hog louse
Haematopota crassicornis Wahlberg } *Haematopota pluvialis* (L.) }	common clegs
Haemolaelaps casalis (Berlese), see *Androlaelaps casalis*	
HALTICINAE, see ALTICINAE	
Haplodiplosis equestris (Wagner), see *Haplodiplosis marginata*	
Haplodiplosis marginata (von Roser)	saddle gall midge
Harmandia loewi (Rübsaamen)	poplar leaf gall midge
Harmandia tremulae (Winnertz)	aspen leaf gall midge
HARPALINI (tribe within the subfamily Carabinae)	ground beetles
Harpalus spp.	black-lustred ground beetles
Harpalus rufipes (Degeer)	strawberry seed beetle
Hartigiola annulipes (Hartig)	beech pouch-gall midge
see also *Mikiola fagi*	
Hauptidia maroccana (Melichar)	glasshouse leafhopper
Hayhurstia atriplicis (L.)	chenopod aphid

35

SCIENTIFIC NAMES—Arthropods

Scientific Name	Common Name
Hedya dimidioalba (Retzius)	marbled orchard tortrix (moth) fruit tree tortrix (see also species of *Acleris, Apotomis, Archips, Ditula, Olethreutes* and *Pandemis*) larva = green budworm spotted apple budworm

Hedya nubiferana (Haworth), see *Hedya dimidioalba*
Hedya pruniana (Hübner) .. plum tortrix (moth)
Helicomyia saliciperda (Dufour) .. willow shot-hole midge
see also species of *Rhabdophaga* willow wood midge
**Helicoverpa armigera* (Hübner) .. (moth) larva = Old World bollworm
    African bollworm
    corn earworm
    cotton bollworm
    tomato-worm
    adult = scarce bordered straw moth

Heliothis armigera (Hübner), see *Helicoverpa armigera*
Heliothis obsoleta: not *obsoleta* (Fabricius, 1793), although misidentified as such in Hampson, 1903, and in several later published works, see *Helicoverpa armigera*
Heliothis peltigera (Denis & Schiffermüller) bordered straw moth
Heliothis zea: not *zea* (Boddie, 1857), although misidentified as such in several later published works, see *Helicoverpa armigera*
Heliothrips adonidum Haliday, see *Heliothrips haemorrhoidalis*
Heliothrips haemorrhoidalis (Bouché) .. glasshouse thrips
Helophorus nubilus Fabricius .. wheat shoot beetle
    wheat mud beetle
Helophorus porculus Bedel ⎫
Helophorus rufipes (Bosc d'Antic) ⎭ .. turnip mud beetles
Helophorus rugosus Olivier, see *Helophorus rufipes*
  note: the name *rugosus* has been wrongly used in several published works to describe *Helophorus porculus* Bedel
Helops caeruleus (L.) .. violet willow beetle
†*Hemerobius* spp. .. brown lacewings
see also *Wesmaelius* spp.
**Hemiberlesia lataniae* (Signoret) .. latania scale
Hemichionaspis aspidistrae (Signoret), see *Pinnaspis aspidistrae*
HEMIPTERA .. aphids, bugs, hoppers, etc.
†*Hemisarcoptes malus* (Shimer) .. mussel scale predatory mite
Hemitarsonemus latus (Banks), see *Polyphagotarsonemus latus*
Hemitarsonemus tepidariorum (Warburton) fern mite
Henria psalliotae Wyatt .. mushroom midge
see also *Lestremia cinerea, Mycophila barnesi* and *Mycophila speyeri*
Hepialus humuli (L.) .. ghost swift moth
    ghost moth
Hepialus lupulinus (L.) .. garden swift moth
    common swift moth
Heptamelus ochroleucus (Stephens) .. fern stem sawfly
Hercinothrips bicinctus (Bagnall) .. smilax thrips
Hercinothrips femoralis (Reuter) .. banded glasshouse thrips
    sugar beet thrips
**Heterobostrychus aequalis* (Waterhouse) ⎫
**Heterobostrychus brunneus* (Murray) ⎭ .. (beetle) larvae = boxwood borers
Heterogaster urticae (Fabricius) .. nettle ground bug
Heteromurus nitidus (Templeton) .. moss springtail
†*Heteropelma calcator* (Wesmael) .. pine looper parasite (wasp)
see also *Cratichneumon viator, Dusona oxyacanthae* and *Polytribax arrogans*
Heteropeza pygmaea Winnertz .. mushroom cecid (fly)
**Heteropoda venatoria* (L.) .. banana spider

SCIENTIFIC NAMES—Arthropods

Scientific Name	Common Name
HETEROPTERA	bugs
HEXAPODA, see INSECTA	
Hexomyza schineri (Giraud)	sallow stem galler (fly)
†*Himacerus apterus* (Fabricius)	tree damsel bug
Himacerus lativentris (Boheman), see *Aptus mirmicoides*	
Himacerus mirmicoides (Costa), see *Aptus mirmicoides*	
Hippobosca equina L.	forest fly
Histiostoma spp.	sewage mites
Histiostoma feroniarum (Dufour)	damp mite
Hofmannophila pseudospretella (Stainton)	brown house moth
Hohorstiella lata (Piaget)	pigeon body louse
HOMOPTERA	aphids, hoppers, mealybugs, phylloxeras, psyllids, scale insects, whiteflies
Homotoma ficus (L.)	fig sucker (psyllid)
Hoplia philanthus (Fuessly)	Welsh chafer (beetle)
Hoplocampa brevis (Klug)	pear sawfly
Hoplocampa flava (L.)	plum sawfly
Hoplocampa testudinea (Klug)	apple sawfly
Humerobates rostrolamellatus Grandjean	cherry beetle mite
Hyadaphis coniellum Theobald, see *Hyadaphis foeniculi*	
Hyadaphis foeniculi (Passerini)	fly-honeysuckle aphid / honeysuckle aphid (see also *Hyadaphis passerinii*)
Hyadaphis lonicerae Börner, see *Hyadaphis passerinii*	
Hyadaphis passerinii (del Guercio)	honeysuckle aphid
*Hyalomma aegyptium (L.)	tortoise tick
Hyalopteroides dactylidis (Hayhurst), see *Hyalopteroides humilis*	
Hyalopteroides humilis (Walker)	cocksfoot aphid
Hyalopterus amygdali (Blanchard)	mealy peach aphid
Hyalopterus arundinis: description of *arundinis* (Fabricius, 1775) is unresolved; in later published works the name has been applied to *Hyalopterus pruni* (see below)	
Hyalopterus pruni (Geoffroy)	mealy plum aphid
Hybernia aurantiaria (Hübner), see *Agriopis aurantiaria*	
Hybernia defoliaria (Clerck), see *Erannis defoliaria*	
Hydraecia micacea (Esper)	rosy rustic moth / *larva* = potato stem borer / potato skin borer
Hydrellia spp	(fly) *larvae* = cereal leaf miners
see also species of *Agromyza* and *Phytomyza nigra*	
Hydrocampa nympheata (L.), see *Elophila nymphaeata*—note also misspelling	
Hydromya dorsalis (Fabricius)	snail-killing fly
Hydrotaea spp.	sweat flies / cluster flies / swarming flies } see also *Eudasyphora* spp., *Ophyra* spp., *Pollenia* spp. and *Thaumatomya* spp.
see also *Morellia simplex*	
Hydrotaea irritans (Fallén)	sheep head fly / plantation fly
Hydrotaea albipuncta (Zetterstedt) ⎫	
Hydrotaea meteorica (L.) ⎬	common sweat flies
Hydrotaea occulta (Meigen) ⎭	
Hylaea fasciaria (L.)	barred red moth
Hylastes ater (Paykull): attribution to author incorrect, see *Hylastes ater* (Fabricius)	

SCIENTIFIC NAMES—Arthropods

Scientific Name	Common Name
Hylastes angustatus (Herbst)	
Hylastes ater (Fabricius)	
Hylastes attenuatus Erichson	black pine beetles
Hylastes brunneus Erichson	
Hylastes opacus Erichson	
Hylastes cunicularius Erichson	black spruce beetle
Hylastinus obscurus (Marsham)	large broom bark beetle
Hylecoetus dermestoides (L.)	(beetle) *larva* = large timberworm

    Hylemya antiqua (Meigen), see *Delia antiqua*
    Hylemya brassicae (Bouché), see *Delia radicum*
    Hylemya brunnescens: not *brunnescens* (Zetterstedt, 1845), although misidentified as such in several later published works, see *Delia cardui*
    Hylemya cilicrura (Rondani), see *Delia platura*
    Hylemya coarctata (Fallén), see *Delia coarctata*
    Hylemya echinata (Séguy), see *Delia echinata*
    Hylemyia: misspelling, see *Hylemya*
    Hylesinus fraxini: not *fraxini* (Panzer, 1799), although misidentified as such in several later published works, see *Leperisinus varius*

Hylesinus oleiperda (Fabricius)	lesser ash bark beetle
Hylobius abietis (L.)	pine weevil
Hylotrupes bajulus (L.)	house longhorn beetle
Hylurgops palliatus (Gyllenhal)	pine bark beetle

    see also *Pityogenes chalcographus*

HYMENOPTERA	ants, bees, sawflies, wasps, etc.

    Hypera austriaca (Schrank), see *Hypera punctata*
    note: a questionable synonym of *Hypera punctata* (Fabricius)

Hypera nigrirostris (Fabricius)	clover leaf weevils
Hypera postica (Gyllenhal)	lucerne weevils
Hypera punctata (Fabricius)	trefoil leaf weevil

    Hypera variabilis (Herbst), see *Hypera postica*
    Hypera zoilus (Scopoli), see *Hypera punctata*
    note: a questionable synonym of *Hypera punctata* (Fabricius)

Hyperetes, subgenus of *Cerobasis*
Hyperomyzella, subgenus of *Hyperomyzus*

Hyperomyzus lactucae (L.)	currant—sowthistle aphid currant—lettuce aphid (see also *Nasonovia ribisnigri* and *Hyperomyzus pallidus*)
Hyperomyzus pallidus Hille Ris Lambers	gooseberry—sowthistle aphid currant—lettuce aphid (see also *Nasonovia ribisnigri* and *Hyperomyzus lactucae*)
Hyperomyzus rhinanthi (Schouteden)	currant—yellowrattle aphid

    Hyperomyzus tulipaella (Theobald): *tulipaella* (Theobald) is a species synonym, but the name *tulipaellus* is retained for a subspecies, see *Rhopalosiphoninus staphyleae*

Hyphantria cunea (Drury)	(moth) *larva* = fall webworm *adult* = American white moth mulberry moth
Hypnoidus riparius (Fabricius)	bank click beetle *larva* = bank wireworm
Hypoderma bovis (L.)	ox warble flies
Hypoderma lineatum (Villers)	ox bot flies warble flies
Hypoderma diana Brauer	deer warble fly
Hypogastrura armata (Nicolet)	gunpowder-mites
Hypogastrura denticulata (Bagnall)	mushroom springtails (see also *Xenylla* spp.)
Hypogastrura manubrialis (Tullberg)	fine gunpowder-mite (springtail)
Hypokopelates antalus (Hopffer)	brown playboy butterfly

    Hyponomeuta, see *Yponomeuta*

SCIENTIFIC NAMES—Arthropods

Scientific Name	Common Name
HYPONOMEUTIDAE, see YPONOMEUTIDAE	
Hypothenemus hampei (Ferrari)	(beetle) *larva* = coffee berry borer
Hypsopygia costalis (Fabricius)	gold fringe moth *larva* = hayworm
Hystrichopsylla talpae (Curtis)	mole flea
*Icerya purchasi Maskell	fluted scale cottony cushion scale
ICHNEUMONIDAE	ichneumons sail wasps
Idiopterus nephrelepidis Davis	fern aphid
INSECTA	insects
Ips acuminatus (Gyllenhal) see also *Orthotomicus laricis*	pattern engraver beetle
*Ips amitinus Eichhoff	smaller eight-toothed spruce bark beetle
Ips cembrae (Heer)	larch bark beetle
*Ips duplicatus (Sahlberg)	northern spruce bark beetle
Ips sexdentatus (Börner)	six-toothed pine bark beetle
Ips typographus (L.)	spruce bark beetle
Iridomyrmex humilis (Mayr)	Argentine ant
Iridothrips iridis (Watson), see *Frankliniella iridis*	
ISCHNOCERA	chewing lice
see also AMBLYCERA	biting lice (see also RHYNCHOPH-THIRINA)
Ischnodemus sabuleti (Fallén)	European chinch bug
ISOPODA	woodlice sowbugs
*Isotomus speciosus (Schneider)	Austrian longhorn beetle
Ixodes autumnalis: not *autumnalis* Leach, 1815, although misidentified as such in several later published works, see *Ixodes hexagonus*	
Ixodes canisuga Johnston	dog tick
Ixodes hexagonus Leach	hedgehog tick
Ixodes lividus Koch	sand martin tick
Ixodes putus (Cambridge), see *Ixodes uriae*	
Ixodes ricinus (L.)	sheep tick castor bean tick
Ixodes tenuirostris Neumann, see *Ixodes trianguliceps*	
Ixodes trianguliceps Birula	vole and shrew tick
Ixodes uriae White	seabird tick
Ixodes ventalloi Gil Collado	rabbit tick
Ixodes vespertilionis Koch	bat hard tick
IXODIDAE	hard ticks
Jaapiella medicaginis (Rübsaamen)	lucerne leaf midge
Janus luteipes (Lepeletier)	willow stem sawfly
JASSIDAE, see CICADELLIDAE	
Javesella pellucida (Fabricius)	cereal leafhopper
Kakothrips pisivora (Westwood), see *Kakothrips pisivorus*	
Kakothrips pisivorus (Westwood)	pea thrips
Kakothrips robustus (Uzel), see *Kakothrips pisivorus*	
Kaltenbachiola strobi (Winnertz)	spruce cone gall midge
Kermes quercus (L.) see also *Eulecanium ciliatum*	oak scale

SCIENTIFIC NAMES—Arthropods

Scientific Name | Common Name

Kiefferia pericarpiicola (Bremi) .. .. parsnip flower midge
   Kimminsia spp., see *Wesmaelius* spp.
Kissophagus hederae (Schmitt) .. .. ivy bark beetle
   Knemidokoptes gallinae (Railliet), see *Neocnemidocoptes gallinae* (Railliet)
   Knemidokoptes mutans (Robin & Lanquetin) scalyleg mite (of fowls)
   Knemidokoptes pilae Lavoipierre & Griffiths scalyleg mite (of budgerigars)
*Knulliana cinctus (Drury) .. .. .. belted chion beetle
Korscheltellus, subgenus of *Hepialus*

Labia minor (L.) .. .. lesser earwig
Lacanobia oleracea (L.) .. .. .. tomato moth
                                          bright-line brown-eye moth
Lachesilla pedicularia (L.) .. .. .. cosmopolitan grain psocid
   Lachnus rosae Cholodkovsky, see *Maculolachnus submacula*
Laelaps echidninus Berlese .. .. .. spiny rat mite
   Laemophloeus ferrugineus (Stephens), see *Cryptolestes ferrugineus*
   Laemophloeus minutus (Olivier, 1791), not *minutus* (Fourcroy, 1785), see *Cryptolestes pusillus*
Laminosioptes cysticola (Vizioli) .. .. fowl cyst mite
                                          flesh mite
                                          subcutaneous mite
*Lampides boeticus (L.) .. .. .. long-tailed blue butterfly
Lampronia capitella (Clerck) .. .. (moth) *larva* = currant shoot borer
Lampronia capittella (Clerck): misspelling, see entry above
Lampronia rubiella (Bjerkander) .. .. raspberry moth
                                          larva = raspberry borer
                                              raspberry maggot
†*Lampyris noctiluca* (L.) .. .. .. glow-worm (beetle)
Laothoe populi (L.) .. .. poplar hawk moth
   Laphygma exigua (Hübner), see *Spodoptera exigua*
   Laria, see *Bruchus*
   LARVAEVORIDAE, see TACHINIDAE
Lasioderma serricorne (Fabricius) .. .. cigarette beetle
Lasioptera rubi Heeger .. .. .. blackberry stem gall midge
                                          raspberry stem gall midge
Lasius brunneus (Latreille) .. .. .. brown ant
Lasius flavus (Fabricius) .. .. .. yellow meadow ant
                                          mound ant
                                          turf ant
Lasius niger (L.) .. .. common black ant
Lasius mixtus (Nylander) ⎫
Lasius umbratus (Nylander) ⎭ .. .. yellow ants
   Laspeyresia (except entry below), see *Cydia*
   Laspeyresia woeberiana (Denis & Schiffermüller), see *Enarmonia formosana*
Latheticus oryzae Waterhouse .. .. long-headed flour beetle
LATHRIDIIDAE .. .. plaster beetles
   Lecanium corni (Bouché), see *Parthenolecanium corni*
   Lecanium cornicrudum: not *cornicrudum* (Green, 1917), although misidentified as such Green, 1930, and in several later published works, see *Parthenolecanium pomeranicum*
   Lecanium coryli (L.), see *Eulecanium tiliae*
   Lecanium crudum (Green), see *Parthenolecanium corni*
   Lecanium hemisphaericum Targioni-Tozzetti, see *Saissetia coffeae*
   Lecanium hesperidum (L.), see *Coccus hesperidum*
   Lecanium persicae (Fabricius), see *Parthenolecanium persicae*
   Leiognathus sylviarum (Canestrini & Fanzago), see *Ornithonyssus sylviarum*
   Lema melanopa (L.), see *Oulema melanopa*
Leperisinus varius (Fabricius) .. .. ash bark beetle

SCIENTIFIC NAMES—Arthropods

Scientific Name — **Common Name**

Lepidoglyphus destructor (Schrank) .. cosmopolitan food mite
    grocers' itch mite (see also *Glycyphagus domesticus*)
LEPIDOPTERA .. .. butterflies and moths
*Lepidosaphes beckii (Newman) .. .. citrus mussel scale
Lepidosaphes conchyformis (Gmelin in Linnaeus) .. .. fig mussel scale
    Lepidosaphes ficus (Signoret), see *Lepidosaphes conchyformis*
*Lepidosaphes gloverii (Packard) .. .. citrus scale
Lepidosaphes machili (Maskell) .. .. cymbidium scale
Lepidosaphes ulmi (L.) .. .. mussel scale
Lepinotus inquilinus Heyden ⎫
Lepinotus patruelis Pearman ⎬ .. .. black domestic psocids
Lepinotus reticulatus Enderlein ⎭
Lepisma saccharina L. .. .. .. silverfish (bristletail)
    Lepresinus [misspelling of generic name] *fraxini*: not *fraxini* (Panzer, 1799), although misidentified as such in several later published works, see *Leperisinus varius*
†*Leptacis tipulae* (Kirby) .. .. .. wheat blossom midge parasite (wasp)
    Leptideella brevipennis (Mulsant), see *Nathrius brevipennis*
*Leptinotarsa decemlineata (Say) .. .. Colorado beetle
Leptocera caenosa Rondani .. .. .. sewage fly
    Leptohylemyia coarctata (Fallén), see *Delia coarctata*
Leptopsylla segnis (Schoenherr) .. .. house mouse flea
    mouse flea
Lestremia cinerea Macquart .. .. .. mushroom midge
    see also *Henria psalliotae* and species of *Mycophila*
Leucoma salicis (L.) .. .. satin moth
Leucoptera laburnella (Stainton) .. .. (moth) *larva* = laburnum leaf miner
Leucoptera malifoliella (Costa) .. .. pear leaf blister moth
    Leucoptera scitella (Zeller), see *Leucoptera malifoliella*
Lilioceris lilii (Scopoli) .. .. .. lily beetle
    Limothrips avenae Hinds, see *Limothrips cerealium*
Limothrips cerealium Haliday .. .. grain thrips
    corn thrips
    thunderflies
    Linognathus forficulus (Rudow), see *Linognathus stenopsis*
Linognathus ovillus (Neumann) .. .. sheep sucking louse
    face louse
*Linognathus pedalis (Osborn) .. .. sheep foot louse
    Linognathus piliferus (Burmeister), see *Linognathus setosus*
    Linognathus rupicaprae (Rudow), see *Linognathus stenopsis*
    Linognathus schistopygus (Nitzsch), see *Linognathus stenopsis*
Linognathus setosus (von Olfers) .. .. dog sucking louse
Linognathus stenopsis (Burmeister) .. goat sucking louse
    Linognathus tenuirostris (Burmeister), see *Linognathus vituli*
Linognathus vituli (L.) .. .. .. long-nosed cattle louse
Linopodes spp. .. .. long-legged mushroom mites
    (in particular *Linopodes motatorius*)
Liosomaphis berberidis (Kaltenbach) .. barberry aphid
    berberis aphid
Liothrips vaneeckei Priesner .. .. .. lily thrips
    lily bulb thrips
Lipara lucens Meigen .. .. reed gall fly
    Liparis monacha (L.), see *Lymantria monacha*
Lipeurus caponis (L.) .. .. .. chicken wing louse
    Lipeurus heterographus Nitzsch, see *Cuclotogaster heterographus*
    Lipeurus variabilis Burmeister, see *Lipeurus caponis*
    Liponyssus bacoti (Hirst), see *Ornithonyssus bacoti*
    Liponyssus sylviarum (Canestrini & Fanzago), see *Ornithonyssus sylviarum*

SCIENTIFIC NAMES—Arthropods

Scientific Name	Common Name
Lipoptena cervi (L.)	deer fly
Liposcelis bostrychophilus Badonnel	stored product psocid booklouse (see also *Liposcelis subfuscus*)

Liposcelis divinatorius: description of *divinatorius* (Müller, 1776) is unresolved; in later published works the name has been applied to several different species within the genus *Liposcelis*

Liposcelis subfuscus Broadhead	outhouse psocid booklouse (see also *Liposcelis bostrychophilus*)
Liposthenus latreillei (Kieffer)	ivy gall cynipid (wasp)
Liriomyza bryoniae (Kaltenbach)	(fly) *larva* = tomato leaf miner
Liriomyza congesta (Becker) ⎫ *Liriomyza pisivora* Hering ⎭	(fly) *larvae* = pea leaf miners

Liriomyza solani Hering, see *Liriomyza bryoniae*

*Liriomyza trifolii (Burgess in Comstock)	(fly) *larva* = American serpentine leaf miner
Listrophorus gibbus Pagenstecher	rabbit fur mite

see also *Cheyletiella parasitivorax*

LITHOBIIDAE	garden centipedes

(in particular *Lithobius* spp.)
see also GEOPHILIDAE

†*Lithobius forficatus* (L.)	common garden centipede

Lithocolletis coryli Nicelli, see *Phyllonorycter coryli*

†*Litomastix aretas* (Walker)	strawberry tortrix parasite (wasp)
Livia juncorum (Latreille)	rush sucker (psyllid)
*Lobesia botrana (Denis & Schiffermüller)	European vine moth
Lochmaea suturalis (Thomson)	heather beetle
Lomaspilis marginata (L.)	clouded border moth

Longitarsus ater: not *ater* (Fabricius, 1775), although misidentified as such in Fowler, 1890, and in several later published works, see *Longitarsus parvulus*

Longitarsus ferrugineus (Foudras)	mint flea beetle

Longitarsus laevis: not *laevis* (Duftschmid, 1825), although misidentified as such in Allard, 1860, and in several later published works, see *Longitarsus succineus*

Longitarsus parvulus (Paykull)	flax flea beetle linseed flea beetle
Longitarsus succineus (Foudras)	chrysanthemum flea beetle

Longitarsus waterhousei Kutschera, see *Longitarsus ferrugineus*
Longiunguis pyrarius (Passerini), see *Melanaphis pyraria*

*Lophocateres pusillus (Klug)	Siamese grain beetle

Lophyrus pini (L.), see *Diprion pini*
Lophyrus sertiferus (Fourcroy), see *Neodiprion sertifer*
Loxostege sticticalis (L.), see *Margaritia sticticalis*
Loxotropa tritoma (Thomson), see *Basalys tritoma*

LUCANIDAE	stag beetles

(in particular *Lucanus cervus*)

Lucanus cervus (L.)	stag beetle
Lucilia spp.	greenbottles (flies)
Lucilia sericata (Meigen)	sheep maggot fly greenbottle sheep blow-fly
Luperina testacea (Denis & Schiffermüller)	flounced rustic moth
Lycia hirtaria (Clerck)	brindled beauty moth
Lycophotia porphyrea (Denis & Schiffermüller)	true lovers knot moth
Lycoriella auripila (Winnertz) ⎫ *Lycoriella solani* (Winnertz) ⎭	mushroom sciarid flies

see also *Bradysia brunnipes* (Meigen)

LYCTIDAE	powder-post beetles

(in particular *Lyctus* spp. and *Minthea* spp.)

SCIENTIFIC NAMES—Arthropods

Scientific Name	Common Name
Lyctocoris campestris (Fabricius)	stack bug
	debris bug
Lyctus spp.	powder-post beetles
see also *Minthea* spp.	
*Lyctus africanus Lesne	African lyctid (beetle)
Lyctus planicollis LeConte	European powder-post beetle
Lygaeonematus erichsonii (Hartig), see *Pristiphora erichsonii*	
Lygaeonematus laricis (Hartig), see *Pristiphora laricis*	
Lygocoris pabulinus (L.)	common green capsid (bug)
Lygus pabulinus (L.), see *Lygocoris pabulinus*	
Lygus pratensis: not *pratensis* (Linnaeus, 1758), although misidentified as such in several later published works, see *Lygus rugulipennis*	
Lygus rugulipennis Poppius	tarnished plant bug
	bishop bug
‡*Lymantria dispar* (L.)	gypsy moth
Lymantria monacha (L.)	black arches moth
	black-arched tussock
	nun moth
LYMANTRIIDAE	tussock moths
LYMEXYLIDAE	(beetle) *larvae* = timberworms
Lymexylon navale (L.)	(beetle) *larva* = ship timberworm
Lyonetia clerkella (L.)	(moth) *larva* = apple leaf miner
Lyperosia irritans (L.), see *Haematobia irritans*	
Lytta vesicatoria (L.)	blister beetle
	Spanish 'fly'

Scientific Name	Common Name
Macrocheles muscaedomesticae (Scopoli)	house fly mite
Macrodiplosis spp.	oak fold-gall midges
Macropsis fuscula (Zetterstedt) } *Macropsis scotti* Edwards	rubus leafhoppers
Macrosiphoniella sanborni (Gillette)	chrysanthemum aphid
Macrosiphum albifrons Essig	lupin aphid
Macrosiphum avenae (Fabricius), see *Sitobion avenae*	
Macrosiphum dirhodum (Walker), see *Metopolophium dirhodum*	
Macrosiphum euphorbiae (Thomas)	potato aphid
Macrosiphum fragariae (Walker), see *Sitobion fragariae*	
Macrosiphum funestum (Macchiati)	scarce blackberry aphid
Macrosiphum gei: not *gei* (Koch, 1855), although misidentified as such in several later published works, see *Macrosiphum euphorbiae*	
Macrosiphum granarium (Kirby), see *Sitobion avenae*	
Macrosiphum onobrychis (Boyer de Fonscolombe), see *Acyrthosiphon pisum*	
Macrosiphum pelargonii (Kaltenbach), see *Acyrthosiphon malvae*	
Macrosiphum pisi (Kaltenbach), see *Acyrthosiphon pisum*	
Macrosiphum rosae (L.)	rose aphid
Macrosiphum rubiellum Theobald, see *Sitobion fragariae*	
Macrosiphum rubifolium Theobald, see *Macrosiphum funestum*	
Macrosiphum solanifolii Ashmead see *Macrosiphum euphorbiae*	
Maculolachnus submacula (Walker)	rose root aphid
Magdalis armigera (Fourcroy)	apple foliage weevil
Magdalis barbicornis (Latreille)	pear weevil
Malacosoma neustria (L.)	lackey moth
*Mallodon downesi Hope	Ethiopian black longhorn beetle
MALLOPHAGA, see under AMBLYCERA, ISCHNOCERA and RHYNCHOPHTHIRINA	
Mamestra albidilinea (Haworth), see *Mamestra brassicae*	
Mamestra brassicae (L.)	cabbage moth
Mamestra oleracea (L.), see *Lacanobia oleracea*	
Mamestra persicariae (L.), see *Melanchra persicariae*	

SCIENTIFIC NAMES—Arthropods

Scientific Name	Common Name
Margaritia sticticalis (L.)	moth *(larva)* = beet webworm
MARGARODIDAE	scale insects
see also COCCIDAE and DIASPIDIDAE	
Matsucoccus pini (Green)	pine needle scale
Mayetiola avenae (Marchal)	oat stem midge
Mayetiola destructor (Say)	hessian fly
Mecinus pyraster (Herbst)	lesser apple foliage weevil
MECOPTERA	scorpion-flies
Megachile spp.	leaf cutter bees
Megachile centuncularis (L.)	common leaf cutter bee

Megaselia bovista (Gimmerthal) } *Megaselia halterata* (Wood) } *Megaselia nigra* (Meigen) }	mushroom scuttle flies
	  mushroom flies
	  mushroom phorids
	  Worthing phorid (applies only to *M. halterata*)

Megastigmus spp.	seed wasps
Megastigmus spermotrophus Wachtl	Douglas fir seed wasp Douglas fir seedfly
Megastigmus pinus Parfitt	silver fir seed wasp
Megempleurus porculus (Bedel), see *Helophorus porculus*	
Megempleurus rugosus (Olivier), see *Helophorus rufipes*	
Megninia cubitalis (Mégnin)	chicken feather mite feather mite
Megoura viciae Buckton	vetch aphid
Melanagromyza simplex (Loew), see *Ophiomyia simplex*	
Melanaphis pyraria (Passerini)	pear—grass aphid
Melanchra brassicae (L.), see *Mamestra brassicae*	
Melanchra persicariae (L.)	dot moth
MELANDRYIDAE	dry wood boring beetles
Melanoxantherium salicis (L.), see *Pterocomma salicis*	
Melasoma populi (L.), see *Chrysomela populi*	
Meligethes spp.	blossom beetles pollen beetles
Meligethes aeneus (Fabricius) } *Meligethes viridescens* (Fabricius) }	bronzed blossom beetles
Meliphora grisella (Fabricius), see *Achroia grisella*	
Meloe spp.	oil beetles
Melolontha hippocastani Fabricius } *Melolontha melolontha* (L.) }	cockchafers (beetles) Maybugs *larvae* = white grubs
Melolontha vulgaris Fabricius, see *Melolontha melolontha*	
Melophagus ovinus (L.)	sheep ked (fly)
Menacanthus cornutus (Schömmer) } *Menacanthus pallidulus* (Neumann) }	lesser chicken body lice
Menacanthus meleagridis: not *meleagridis* (Linnaeus, 1758), although misidentified as such in Panzer, 1793, and in several later published works, see *Menacanthus stramineus*	
Menacanthus stramineus (Nitzsch)	chicken body louse broad body louse
Menopon gallinae (L.)	chicken shaft louse
Menopon pallidum (Nitzsch), see *Menopon gallinae*	
Menopon trigonocephalum (von Olfers), see *Menopon gallinae*	
Merodon equestris (Fabricius)	large narcissus fly bulb fly
Meromyza saltatrix (L.)	grass fly
Mesapamea secalis (L.)	common rustic moth
Mesographe forficalis (L.), see *Evergestis forficalis*	
Mesoligia literosa (Haworth)	rosy minor moth

SCIENTIFIC NAMES—Arthropods

Scientific Name   **Common Name**

Metallus albipes (Cameron) ⎫
Metallus pumilus (Klug) ⎭   raspberry leaf mining sawflies
Metallus gei (Brischke)   geum sawfly
  see also *Monophadnoides geniculatus*
Metatetranychus pilosus (Canestrini & Fanzago), see *Panonychus ulmi*
Metatetranychus ulmi (Koch), see *Panonychus ulmi*
Metopolophium dirhodum (Walker)   rose-grain aphid
Metopolophium festucae (Theobald)   fescue aphid
                                                 grass aphid
*Mezium americanum (Laporte de Castelnau)   American spider beetle
  Miana literosa (Haworth), see *Mesoligia literosa*
  Miana strigilis (Clerck): attribution to author incorrect, see *Oligia strigilis* (Linnaeus)
Miarus campanulae (L.)   bellflower gall weevil
Micropeza corrigiolata (L.)   stilt-legged fly
Mikiola fagi (Hartig)   beech pouch-gall midge
  see also *Hartigiola annulipes*
Mindarus abietinus Koch   balsam twig aphid
Mindarus obliquus (Cholodkovsky)   spruce twig aphid
*Minthea spp.   powder-post beetles
  see also *Lyctus* spp.
MIRIDAE   capsid bugs
Monalocoris filicis (L.)   bracken bug
  see also *Neotrombicula autumnalis*   fern bug
Monarthropalpus buxi (Laboulbène)   box leaf mining midge
*Monochamus spp.   sawyer beetles
Monomorium pharaonis (L.)   Pharaoh's ant
Monophadnoides geniculatus (Hartig)   geum sawfly
  see also *Metallus gei*
Monopis rusticella (Hübner)   skin moth
Morellia simplex (Loew)   cattle sweat fly
                                     sweat fly (see also *Hydrotaea* spp.)
Moritziella corticalis (Kaltenbach)   oak bark phylloxera
Musca autumnalis Degeer   face fly
                                   autumn fly
Musca domestica L.   house fly
Muscina stabulans (Fallén)   false stable fly
MUTILLIDAE   velvet ants
Mycetaea hirta (Marsham)   hairy fungus beetle
  see also *Typhaea stercorea*   hairy cellar beetle
MYCETOPHAGIDAE   fungus beetles
MYCETOPHILIDAE   fungus flies
                                     fungus gnats
                                     larvae = cucumber root maggots
                                                   dung maggots
                                                   fungus maggots
Mycophila barnesi Edwards ⎫
Mycophila speyeri (Barnes) ⎭   mushroom midges
  see also *Henria psalliotae* and *Lestremia cinerea*
  Myelois ceratoniae Zeller, see *Ectomyelois ceratoniae*
  Myelophilus piniperda (L.), see *Tomicus piniperda*
Myocoptes musculinus (Koch)   myocoptic mange mite
  MYRIAPODA, see DIPLOPODA and CHILOPODA
Myrmica rubra (L.) ⎫
Myrmica ruginodis Nylander ⎭   red ants
  Mytilaspis pomorum (Bouché), see *Lepidosaphes ulmi*
  Mytilococcus ulmi (L.), see *Lepidosaphes ulmi*
Myzaphis rosarum (Kaltenbach)   lesser rose aphid
Myzocallis coryli (Goeze)   hazel aphid
  Myzodes ligustri (Kaltenbach), see *Myzus ligustri*

SCIENTIFIC NAMES—Arthropods

Scientific Name   Common Name

Myzosiphon, subgenus of *Rhopalosiphoninus*
Myzus ascalonicus Doncaster   shallot aphid
Myzus cerasi (Fabricius)   cherry blackfly
   Myzus circumflexus (Buckton), see *Aulacorthum circumflexum*
   Myzus festucae Theobald, see *Metopolophium festucae*
   Myzus hieracii: not *hieracii* (Kaltenbach, 1843), although misidentified as such in several later published works, see *Nasonovia ribisnigri*
   Myzus kaltenbachi (Schouteden), see *Nasonovia ribisnigri*
   Myzus lactucae (Schrank), see *Nasonovia ribisnigri*
Myzus ligustri (Mosley)   privet aphid
Myzus ornatus Laing   violet aphid
Myzus persicae (Sulzer)   peach—potato aphid
   Myzus pseudosolani Theobald, see *Aulacorthum solani*

   Nabis apterus (Fabricius), see *Himacerus apterus*
Nacerdes melanura (L.)   (beetle) *larva* = wharf borer
Naenia typica (L.)   gothic moth
Nanna spp.   timothy flies
Napomyza carotae Spencer   (fly) *larva* = carrot miner
Napomyza lateralis (Fallén)   calendula fly
   Napomyza populi (Kaltenbach), see *Paraphytomyza populi*
   Nasonovia ribicola (Kaltenbach), see *Nasonovia ribisnigri*
Nasonovia ribisnigri (Mosley)   currant-lettuce aphid
   see also *Hyperomyzus lactucae* and *H. pallidus*   lettuce aphid
Nathrius brevipennis (Mulsant)   wicker longhorn beetle
   see also *Gracilia minuta*
†*Nebria brevicollis* (Fabricius)   common black ground beetle
Necrobia ruficollis (Fabricius)   red-shouldered ham beetle
Necrobia rufipes (Degeer)   copra beetle
         red-legged ham beetle
   Necrophorus: misspelling, see *Nicrophorus*
Nectarosiphon, subgenus of *Myzus*
Nemapogon cloacella (Haworth)   cork moth
Nemapogon granella (L.)   corn moth
         European grain moth
         wolf moth
Nematus leucotrochus Hartig   pale spotted gooseberry sawfly
Nematus melanaspis Hartig   gregarious poplar sawfly
Nematus olfaciens Benson   black currant sawfly
Nematus pavidus Lepeletier   lesser willow sawfly
   Nematus proximus Lepeletier, see *Pontania proxima*
Nematus ribesii (Scopoli)   common gooseberry sawfly
Nematus salicis (L.)   willow sawfly
Nematus spiraeae Zaddach & Brischke   spiraea sawfly
   Nematus viminalis (L.), see *Pontania viminalis*
   Neochmosis cupressi (Buckton), see *Cinara cupressi*
   Neochmosis tujae (del Guercio), see *Cinara cupressi*
Neoclytus acuminatus (Fabricius)   (beetle) *larva* = red-headed ash borer
Neoclytus caprea (Say)   (beetle) *larva* = banded ash borer
Neocnemidocoptes gallinae (Railliet)   depluming mite
         depluming itch mite
   Neodiprion rufus Latreille, see *Neodiprion sertifer*
Neodiprion sertifer (Fourcroy)   fox-coloured sawfly
         lesser pine sawfly
         small pine sawfly
   Neomyzaphis abietina (Walker), see *Elatobium abietinum*

SCIENTIFIC NAMES—Arthropods

Scientific Name	Common Name
Neomyzus, subgenus of *Aulacorthum*	
Neotrombicula autumnalis (Shaw)	harvest mite
	bracken-bug (see also *Monalocoris filicis*)
Nephrotoma spp.	spotted crane flies
	tiger crane flies
Nepticula anomalella (Goeze), see *Stigmella anomalella*	
Nepticula malella (Stainton), see *Stigmella malella*	
NEUROPTERA	lacewings, etc.
Neuroterus albipes (Schenck)	oak leaf smooth-gall cynipid (wasp)
	Schenck's gall wasp
	smooth spangle gall wasp
Neuroterus lenticularis (Olivier), see *Neuroterus quercusbaccarum*	
Neuroterus numismalis (Fourcroy)	oak leaf blister-gall cynipid (wasp)
Neuroterus quercusbaccarum (L.)	oak leaf spangle-gall cynipid (wasp)
Neuroterus tricolor (Hartig)	oak leaf cupped-gall cynipid (wasp)
Neurotoma flaviventris (Retzius), see *Neurotoma saltuum*	
Neurotoma saltuum (L.)	social pear sawfly
Nezara viridula (L.)	green vegetable bug
Nicrophorus spp.	burying beetles
	sexton beetles
Niditinea fuscella (L.)	brown-dotted clothes moth
Niditinea fuscipunctella (Haworth), see *Niditinea fuscella*	
Niptus hololeucus (Faldermann)	golden spider beetle
Noctua comes (Hübner)	lesser yellow underwing moth
Noctua pronuba (L.)	large yellow underwing moth
	larva = cutworm
	surface caterpillar } see also *Agrotis* spp. and *Euxoa* spp.
Nosopsyllus fasciatus (Bosc)	rat flea
†*Notiophilus biguttatus* (Fabricius)	burnished ground beetle
Notocelia, subgenus of *Epiblema*	
Notocelia uddmanniana (L.), see *Epiblema uddmanniana*	
Notoedres cati (Hering)	cat head mange mite
	notoedric itch mite
Notoedres cuniculi (Gerlach), see *Notoedres cati*	
Notoedres muris (Mégnin)	rat ear mange mite
	rat head itch mite
	rat head mange mite
Numonia suavella (Zincken)	porphyry knot-horn moth
Nygmia phaeorrhoea (Donovan), see *Euproctis chrysorrhoea*	
Nygmia phaeorrhoeus (Haworth), see *Euproctis chrysorrhoea*	
Nymphopsocus destructor Enderlein, see *Psyllipsocus ramburii*	
Nymphula nymphaeata (L.), see *Elophila nymphaeata*	
Ochlerotatus, subgenus of *Aedes*	
ODONATA	dragonflies, damselflies
Odontopera bidentata (Clerck)	scalloped hazel moth
Oeciacus hirundinis (Jenyns), see *Oeciacus hirundinis* (Lamark)	
Oeciacus hirundinis (Lamark)	martin bug
	swallow bug
Oestrus ovis L.	sheep nostril fly
	sheep bot fly
Oinophila v-flava (Haworth)	yellow v moth
Olethreutes spp.	fruit tree tortrices (moths)
see also species of *Acleris, Apotomis, Archips, Ditula, Hedya* and *Pandemis*	
Olethreutes lacunana (Denis & Schiffermüller)	dark strawberry tortrix (moth)

SCIENTIFIC NAMES—Arthropods

Scientific Name	Common Name
Oligia strigilis (L.)	marbled minor moth
Oligonychus ulmi (Koch), see *Panonychus ulmi*	
Oligonychus ununguis (Jacobi)	conifer spinning mite
	spruce mite
	spruce spider mite
†*Oligota flavicornis* (Boisduval & Lacordaire)	minute predatory rove beetle
Oligotrophus juniperinus (L.)	juniper gall midge
	whooping gall midge
Omocestus viridulus (L.)	common green grasshopper
Oniscus asellus L.	grey garden woodlouse
Onychiurus spp.	white blind springtails
	ground-fleas
Operophtera brumata (L.)	winter moth
	small winter moth
Operophtera fagata (Scharfenberg)	northern winter moth
	beech winter moth
Ophiomyia simplex (Loew)	(fly) *larva* = asparagus miner
†*Ophion* spp.	red ichneumons (wasps)
†*Ophion luteus* (L.)	yellow ophion
Ophionyssus natricis (Gervais)	snake mite
Ophonus pubescens (Müller), see *Harpalus rufipes*	
Ophyra spp.	cluster flies
see also *Eudasyphora* spp., *Hydrotaea* spp., *Pollenia* spp. and *Thaumatomyia* spp.	swarming flies
Ophyra ignava (Harris)	poultry-house fly
Ophyra leucostoma (Wiedemann), see *Ophyra ignava*	
OPILIONES	harvestmen
	harvesters
	harvest-spiders
*Opogona sacchari (Bojer)	(moth) *larva* = sugar can borer
Opomyza spp.	grass and cereal flies
see also *Geomyza tripunctata*	
Opomyza florum (Fabricius)	yellow cereal fly
Opomyza germinationis (L.)	dusky-winged cereal fly
Orchesella spp.	hairy ground springtails
	orchid springtails
Orchestes fagi (L.), see *Rhynchaenus fagi*	
Orchopeas howardi (Baker)	grey squirrel flea
Orchopeas wickhami (Baker), see *Orchopeas howardi*	
Orgyia antiqua (L.)	vapourer moth
	rusty tussock
Oria musculosa (Hübner)	Brighton wainscot moth
*Orius insidiosus (Say)	insidious flower bug
Orius minutus: not *minutus* (Linnaeus, 1758), although misidentified as such in several later published works, see *Orius vicinus*	
†*Orius vicinus* (Ribaut)	raspberry bug
ORIBATEI, see CRYPTOSTIGMATA	
ORIBATIDAE, see CRYPTOSTIGMATA	
Orneodes hexadactyla (L.), see *Aluctia hexadactyla*	
Ornithomya spp.	bird parasitic flies
Ornithonyssus bacoti (Hirst)	tropical rat mite
Ornithonyssus sylviarum (Canestrini & Fanzago)	northern fowl mite
	European fowl mite
Orthezia insignis Browne	glasshouse orthezia (Homoptera)
*Orthomorpha coarctata (Saussure), see *Asiomorpha coarctata*	
ORTHOPTERA, see SALTATORIA	
Orthosia gothica (L.)	Hebrew character (moth)

SCIENTIFIC NAMES—Arthropods

Scientific Name	Common Name
Orthosia gracilis (Denis & Schiffermüller)	powdered quaker moth
Orthosia incerta (Hufnagel)	clouded drab moth
Orthotomicus laricis (Fabricius)	pattern engraver beetle
see also *Ips acuminatus*	
†*Orthotylus marginalis* Reuter	dark green apple capsid (bug)
†*Orthotylus nassatus* (Fabricius)	lime capsid (bug)
Oryzaephilus mercator (Fauvel)	merchant grain beetle
Oryzaephilus surinamensis (L.)	saw-toothed grain beetle
Oscinella frit (L.)	frit fly
Osmia rufa (L.)	mason bee
Ostrinia nubilalis (Hübner)	(moth) *larva* = European corn borer
Otiorhynchus spp.	wingless weevils
Otiorhynchus clavipes (Bonsdorff)	red-legged weevil
	plum weevil
**Otiorhynchus crataegi* Germar	Mediterranean hawthorn weevil
Otiorhynchus ovatus (L.)	strawberry weevil
	strawberry root weevil (see also *Otiorhynchus rugifrons, Otiorhynchus rugosostriatus* and *Sciaphilus asperatus*)
Otiorhynchus rugifrons (Gyllenhal) }	lesser strawberry weevils
Otiorhynchus rugosostriatus (Goeze) }	strawberry root weevils (see also *Otiorhynchus ovatus* and *Sciaphilus asperatus*)
Otiorhynchus picipes (Fabricius), see *Otiorhynchus singularis*	
Otiorhynchus singularis (L.)	clay-coloured weevil
Otiorhynchus sulcatus (Fabricius)	vine weevil
	black vine weevil
Otiorrhynchus: misspelling, see *Otiorhynchus*	
Otodectes cynotis (Hering)	ear mange mite
	cat ear mite
	dog ear mite
	ear canker mite
	ear mite
Otonyssus autumnalis (Shaw), see *Neotrombicula autumnalis*	
Oulema melanopa (L.)	cereal leaf beetle
	oat leaf beetle
Oxidus gracilis (Koch)	glasshouse millepede
Oxylipeurus polytrapezius (Burmeister)	turkey wing louse
Oxytelus spp., see *Anotylus* spp.	
**Pachydissus hector* Kolbe	Tanzanian coffee longhorn beetle
Pachypappa tremulae (L.)	spruce root aphid
Pachyprotasis variegata (Fallen)	potato sawfly
Pachynematus imperfectus (Zaddach and Brischke)	larch sawfly
see also *Pristiphora wesmaeli*	
Pachyrhina maculata (Meigen), see *Nephrotoma appendiculata*	
**Palorus ratzeburgi* (Wissmann)	small-eyed flour beetle
Palorus subdepressus (Wollaston)	depressed flour beetle
Pammene rhediella (Clerck)	fruitlet mining tortrix (moth)
Pandemis cerasana (Hübner)	barred fruit-tree tortrix (moth)
Pandemis heparana (Denis & Schiffermüller)	dark fruit-tree tortrix (moth)
	fruit tree tortrix (see also species of *Acleris, Apotomis, Archips, Ditula, Hedya* and *Olethreutes*)
Pandemis ribeana (Hübner), see *Pandemis cerasana*	

SCIENTIFIC NAMES—Arthropods

Scientific Name	Common Name

Panolis flammea (Denis & Schiffermüller) .. pine beauty moth
   Panolis piniperda (Panzer), see *Panolis flammea*
Panonychus ulmi (Koch) .. .. .. fruit tree red spider mite
                                                 European red mite
   Panoplia cruciana (L.), see *Epinotia cruciana*
   Panoplia penkleriana: not *penkleriana* (Denis & Schiffermüller, 1775), although misidentified as such in several later published works, see *Epinotia tenerana*
Panorpa communis (L.) .. .. .. common scorpion-fly
Paraceras melis (Walker) .. .. .. badger flea
   Paradesmus gracilis (Koch), see *Oxidus gracilis*
Paradrina, subgenus of *Caradrina*
Paralipsa gularis (Zeller) .. .. .. stored nut moth
**Paramyelois transitella* (Walker) .. .. (moth) *larva* = navel orangeworm
**Paraphytomyza dianthicola* (Venturi) .. (fly) *larva* = Mediterranean carnation leaf miner
Paraphytomyza populi (Kaltenbach) .. (fly) *larva* = poplar leaf miner
   Paraspiniphora bergenstammii (Mik), see *Spiniphora bergenstammi*
   Paratetranychus pilosus Canestrini & Fanzago, see *Panonychus ulmi*
   Paratetranychus ununguis: Jacobi, not Zachvatkin, see *Oligonychus ununguis*
Parathecabius, subgenus of *Pemphigus*
Paratillus carus (Newman) .. .. .. metallic-blue clerid (beetle)
   Paravespula germanica (Fabricius), see *Vespula germanica*
   Paravespula vulgaris (L.), see *Vespula vulgaris*
Parlatoria proteus (Curtis) .. .. .. cattleya scale
Paroxyna misella (Loew) .. .. .. chrysanthemum stem fly
Parthenolecanium corni (Bouché) .. .. brown scale
Parthenolecanium persicae (Fabricius) .. peach scale
Parthenolecanium pomeranicum (Kawecki) yew scale
Parthenothrips dracaenae (Heeger) .. banded-wing palm thrips
                                                 dracaena thrips
                                               palm thrips (see also *Thrips palmi*)
   Pealius avellanae: not *avellanae* (Signoret, 1868), although misidentified as such in Trehan, 1940, and in several later published works, see *Pealius quercus*
Pealius azaleae (Baker & Moles) .. .. azalea whitefly
                                               small rhododendron whitefly
Pealius quercus (Signoret) .. .. .. oak whitefly
   Pediculoides ventricosus (Newport), see *Pyemotes tritici*
   Pediculopsis graminum (Reuter), see *Siteroptes graminum*
Pediculus capitis Degeer .. .. .. head louse (of man)
   Pediculus corporis Degeer, see *Pediculus humanus*
Pediculus humanus L. .. .. .. body louse (of man)
                                               clothes louse
   Pediculus tabescentium Alt, see *Pediculus humanus*
   Pediculus vestimenti Nitzsch, see *Pediculus humanus*
   Pegohylemyia gnava (Meigen), see *Botanophila gnava*
   Pegomya betae (Curtis), see *Pegomya hyoscyami*
   Pegomya dentiens (Pandellé), see *Pegomya rubivora* (Coquillett)
Pegomya hyoscyami (Panzer) .. .. mangold fly
                                               beet fly
                                               mangel fly
                                               larva = beet leaf miner
   Pegomya hyoscyami v *betae* (Curtis), see *Pegomya hyoscyami* (Panzer)
Pegomya nigritarsis (Zetterstedt) .. .. (fly) *larva* = dock miner
Pegomya rubivora (Coquillett) .. .. loganberry cane fly
   Pegomyia: misspelling, see *Pegomya*
   Pemphigus affinis (Kaltenbach), see *Thecabius affinis*
   Pemphigus auriculae (Murray), see *Thecabius auriculae*
Pemphigus bursarius (L.) .. .. .. lettuce root aphid
                                               poplar—lettuce aphid

SCIENTIFIC NAMES—Arthropods

Scientific Name	Common Name

Pemphigus filaginis (Boyer de Fonscolombe), see *Pemphigus populinigrae*
Pemphigus phenax Börner & Blunck ..   carrot root aphid
Pemphigus populinigrae (Schrank) ..   poplar—cudweed aphid
   Pemphigus spirothecae Passerini: misspelling see *Pemphigus spyrothecae*
Pemphigus spyrothecae Passerini .. ..   poplar spiral-gall aphid
Pentarthrum huttoni Wollaston .. ..   Chilean wood-boring weevil
                                            wood boring weevil (see also species of *Euophryum*)
Pentatoma rufipes (L.) .. .. ..   forest bug
                                            cherry stink bug
                                            place bug
PENTATOMIDAE .. ..   shield bugs
  see also ACANTHOSOMIDAE, CYDNIDAE and SCUTELLERIDAE
Pentatrichopus, subgenus of *Chaetosiphon*
   Pentatrichopus fragaefolii (Cockerell), see *Chaetosiphon fragaefolii*
   Pentatrichopus fragariae (Theobald), see *Chaetosiphon fragaefolii*
Penthaleus major (Dugès) .. .. ..   red-legged earth mite
                                            blue oat mite
                                            pea mite
                                            winter grain mite
Peridroma margaritosa (Haworth), see *Peridroma saucia*
Peridroma porphyrea (Denis & Schiffermüller), see *Lycophotia porphyrea*
   note: the name *porphyrea* was wrongly used in Edelsten, 1939, and in several later published works, to describe *Peridroma saucia* (Hübner)
Peridroma saucia (Hübner) .. .. ..   pearly underwing moth
                                            larva = variegated cutworm
Periphyllus californiensis (Shinji) .. ..   California maple aphid
Periplaneta americana (L.) .. .. ..   American cockroach
                                            ship cockroach
Periplaneta australasiae (Fabricius) ..   Australian cockroach
Periplaneta brunnea Burmeister .. ..   brown cockroach
Peritrechus lundi (Gmelin) .. .. ..   potato ground bug
Peronea, see *Acleris*
Petrobia latens (Müller) .. .. ..   stone mite
                                            brown wheat mite
Petrova resinella (L.) .. .. ..   pine resin-gall moth
Pexicopia malvella (Hübner) .. ..   hollyhock seed moth
Phaedon armoraciae (L.) ⎫
Phaedon cochleariae (Fabricius) ⎬ .. ..   mustard beetles
                                            (*P. cochleariae* also known as watercress beetle)

Phaedon tumidulus (Germar) .. ..   celery leaf beetle
   Phagocarpus permunda (Harris) ⎫
   Phagocarpus purmundus: misspelling ⎬ see *Anomia permunda*
   Phalaena typica L., see *Naenia typica*
Phalangium opilio L. .. .. ..   common harvestman
Phalera bucephala (L.) .. .. ..   buff-tip moth
*Pharaxonotha kirschi Reitter .. ..   Mexican grain beetle
PHASMIDA .. ..   stick insects and leaf insects
Phenacoccus aceris (Signoret) .. ..   apple mealybug
   Phigalia pedaria (Fabricius), see *Apocheima pilosaria*
   Phigalia pilosaria (Denis & Schiffermüller), see *Apocheima pilosaria*
   Philaenus leucophthalmus (L.), see *Philaenus spumarius*
Philaenus spumarius (L.) .. .. ..   common froghopper
                                            nymph = cuckoo-spit bug
   Philedone prodromana (Hübner), see *Philedonides lunana*
Philedonides lunana (Thunberg) .. ..   Walker's heath tortrix (moth)
Philopedon plagiatus (Schaller) .. ..   sand weevil
   Philophylla heraclei (L.), see *Euleia heraclei*

SCIENTIFIC NAMES—Arthropods
Scientific Name   Common Name

Philoscia muscorum (Scopoli) . . . .   hedgerow and grassland woodlouse
Phloeomyzus passerinii (Signoret) . . . .   woolly poplar-stem aphid
Phloeophthorus rhododactylus (Marsham) . .   small broom bark beetle
Phlogophora meticulosa (L.) . . . .   angle-shades moth
*Phoracantha semipunctata (Fabricius) . .   eucalyptus longhorn beetle
                                                     larva = phoracantha borer
   Phorbia brassicae (Bouché), see *Delia radicum*
   Phorbia floralis (Fallén), see *Delia floralis*
   Phorbia genitalis: not *genitalis* (Schnabl & Dziedzicki, 1911), although misidentified as such in several later published works, see *Phorbia securis*
Phorbia securis Tiensuu ⎫
Phorbia sepia (Meigen) ⎭ . . . . . .   late-wheat shoot flies
PHORIDAE . . . .   scuttle flies
                                                     phorid flies
Phormia terraenovae Robineau-Desvoidy . .   blackbottle (fly)
Phorodon humuli (Schrank) . . . . . .   damson—hop aphid
                                                     hop—damson aphid
   Phorodon pruni (Scopoli), see *Phorodon humuli*
*Phryneta leprosa (Fabricius) . . . .   African brown longhorn beetle
†*Phryxe nemea* (Meigen) . . . . . .   cabbage moth parasite (fly)
†*Phryxe vulgaris* (Fallén) . . . . . .   silver y moth parasite (fly)
PHTHIRAPTERA . . . .   biting lice, chewing lice and sucking lice
   Phthirus, see *Pthirus*
   Phthorimaea artemisiella (Treitschke), see *Scrobipalpa artemisiella*
**Phthorimaea operculella* (Zeller) . . . .   potato moth
                                                     potato tuber moth
Phyllaphis fagi (L.) . . . .   beech aphid
Phyllobius spp. . . . .   leaf weevils
Phyllobius argentatus (L.) . . . . . .   silver-green leaf weevil
Phyllobius maculicornis Germar . . . .   green leaf weevil
Phyllobius oblongus (L.) . . . . . .   brown leaf weevil
Phyllobius pomaceus Gyllenhal . . . .   nettle leaf weevil
Phyllobius pyri (L.) ⎫
Phyllobius vespertinus (Fabricius) ⎭ . .   common leaf weevils
Phyllocnistis unipunctella (Stephens) . .   (moth) larva = poplar leaf miner
   Phyllocoptes fockeui Nalepa & Trouessart, see *Aculus fockeui*
Phyllocoptes goniothorax (Nalepa) . . . .   hawthorn leaf erineum mite
Phyllocoptes gracilis (Nalepa) . . . .   raspberry leaf and bud mite
   Phyllocoptes lycopersici Massee, see *Aculops lycopersici*
   Phyllocoptes schlechtendali Nalepa, see *Aculus schlechtendali*
   Phyllodecta cavifrons Thomson, see *Phyllodecta laticollis*
Phyllodecta laticollis Suffrian . . . .   small poplar leaf beetle
Phyllodecta vitellinae (L.) . . . . . .   brassy willow leaf beetle
Phyllodecta vulgatissima (L.) . . . .   blue willow leaf beetle
   Phyllodromia supellectilium (Serville), see *Supella longipalpa*
Phyllonorycter coryli (Nicelli) . . . .   nut leaf blister moth
Phyllonorycter messaniella (Zeller) . .   Zeller's midget moth
Phyllopertha horticola (L.) . . . .   garden chafer (beetle)
                                                     June bug
Phyllotreta spp. . . . .   flea beetles
Phyllotreta aerea Allard . . . . . .   small black flea beetle
Phyllotreta atra (Fabricius) ⎫
Phyllotreta consobrina (Curtis) ⎪
Phyllotreta cruciferae (Goeze) ⎬ . .   turnip flea beetles
Phyllotreta nigripes (Fabricius) ⎭   turnip 'fly'
Phyllotreta diademata Foudras . . . .   crown flea beetle
Phyllotreta nemorum (L.) . . . . . .   large striped flea beetle
   Phyllotreta punctulata: description of *punctulata* (Marsham, 1802) is unresolved; in later published works the name has been applied to *Phyllotreta aerea* (see above)

SCIENTIFIC NAMES—Arthropods

Scientific Name   **Common Name**

Phyllotreta undulata Kutschera   small striped flea beetle
Phyllotreta vittula Redtenbacher   barley flea beetle
Phylloxera glabra (von Heyden)   oak leaf phylloxera
  Phylloxera punctata Lichtenstein, see *Phylloxera glabra*
  Phylloxera salicis (Lichtenstein), see *Phylloxerina salicis*
  Phylloxera vastatrix (Planchon), see *Daktulosphaira vitifoliae*
  Phylloxera vitifolii (Fitch), see *Daktulosphaira vitifoliae*
PHYLLOXERIDAE   phylloxeras
Phylloxerina salicis (Lichtenstein)   willow bark phylloxera
Phymatocera aterrima (Klug)   Solomon's seal sawfly
Phymatodes testaceus (L.)   (beetle) *larva* = tanbark borer
  Phymatodes variabilis (L.), see *Phymatodes testaceus*
Physokermes piceae (Schrank)   spruce whorl scale
  Phytagromyza populi (Kaltenbach), see *Paraphytomyza populi*
Phytobia cambii (Hendel)   (fly) *larva* = poplar and willow cambium miner
Phytobia cerasiferae (*Kangas*)   (fly) *larva* = plum cambium miner
                                          prunus cambium miner
                                        adult = plum cambium midge
Phytodecta decemnotata (Marsham)   aspen leaf beetle
  Phytomatocera: misspelling see *Phymatocera*
Phytomyza aquilegiae Hardy   (fly) *larva* = columbine leaf miner
  see also *Phytomyza minuscula*
  Phytomyza atricornis: not *atricornis* Meigen, 1838; in later published works the name has been applied to two species: *Phytomyza horticola* and *Phytomyza syngenesiae*
Phytomyza horticola Goureau   (fly) *larva* = chrysanthemum leaf miner
  see also *Phytomyza syngenesiae*
Phytomyza ilicis Curtis   (fly) *larva* = holly leaf miner
Phytomyza minuscula Goureau   (fly) *larva* = columbine leaf miner
  see also *Phytomyza aquilegiae*
Phytomyza nigra Meigen   (fly) *larva* = cereal leaf miner
  see also species of *Agromyza* and *Hydrellia* spp.
Phytomyza ramosa Hendel   teasel fly
Phytomyza rufipes Meigen   (fly) *larva* = cabbage leaf miner
Phytomyza syngenesiae (Hardy)   (fly) *larva* = chrysanthemum leaf miner
  see also *Phytomyza horticola*
Phytonemus pallidus (Banks)   cyclamen mite
                                          begonia mite
Phytonemus pallidus
        ssp. *fragariae* (Zimmerman)   strawberry mite
  Phytonomus austriacus (Schrank), see *Hypera punctata*
    note: a questionable synonym of *Hypera punctata* (Fabricius)
  Phytonomus nigrirostris (Fabricius), see *Hypera nigrirostris*
  Phytonomus punctata (Fabricius), see *Hypera punctata*
  Phytonomus variabilis (Herbst), see *Hypera postica*
☆*Phytoptus avellanae* Nalepa   filbert bud mite
                                          nut gall mite
  Phytoptus essigi (Hassan), see *Acalitus essigi*
  Phytoptus fockeui (Nalepa & Trouessart), see *Aculus fockeui*
  Phytoptus gracilis (Nalepa), see *Phyllocoptes gracilis*
  Phytoptus laevis Nalepa, see *Eriophyes laevis*
  Phytoptus padi Nalepa, see Eriophyes padi
  Phytoptus phloeocoptes Nalepa, see *Acalitus phloeocoptes*
  Phytoptus piri Pagenstecher: misspelling, see *Eriophyes pyri*
  Phytoptus psilaspis Nalepa, see *Cecidophyopsis psilaspis*

☆ Note: the use of the generic names *Phytoptus* and *Eriophyes* follows that in general use before 1971, and not that proposed by Newkirk, R. A. and Keifer, H. H., (1971) *Eriophyid Studies* C5, California Department of Agriculture, Sacramento.

SCIENTIFIC NAMES—Arthropods

Scientific Name **Common Name**

Phytoptus pyri Pagenstecher, see *Eriophyes pyri*
Phytoptus ribis Westwood, see *Cecidophyopsis ribis*
Phytoptus rudis Canestrini, see *Acalitus rudis*
Phytoptus similis Nalepa, see *Eriophyes similis*
Phytoptus tiliae Pagenstecher, see *Eriophyes tiliae*
Phytoptus triradiatus Nalepa, see *Eriophyes triradiatus*
Phytoptus vitis Pagenstecher, see *Colomerus vitis*
†*Phytoseiulus persimilis* Athias-Henriot .. two-spotted spider mite predator (mite)
    red spider mite predator
Phytoseiulus riegeli Dosse, see *Phytoseiulus persimilis*
Pieris spp. .. .. cabbage white butterflies
Pieris brassicae (L.) .. .. large white butterfly
Pieris napi (L.) .. .. green-veined white butterfly
Pieris rapae (L.) .. .. small white butterfly
Piesma maculatum (Laporte de Castelnau) spinach beet bug
Piesma quadratum (Fieber) .. .. .. beet leaf bug
    beet bug
    beet lace bug
†*Pimpla hypochondriaca* (Retzius) .. .. red-legged ichneumon (wasp)
Pimpla instigator (Fabricius), see *Pimpla hypochondriaca*
Pimpla pomorum Ratzeburg, see *Scambus pomorum*
Pineus orientalis (Dreyfus) .. .. .. oriental pine adelges
Pineus pineoides (Cholodkovsky) .. .. small spruce adelges
    small spruce woolly aphid
Pineus pini (Macquart) .. .. .. Scots pine adelges
    European pine woolly aphid
  Pineus pini (Gmelin in Linnaeus): attribution to author incorrect, see *Pineus pini* (Macquart)
Pineus similis (Gillette) .. .. .. spruce gall adelges
Pineus strobi (Hartig) .. .. .. Weymouth pine adelges
  Pineus strobus (Ratzeburg), see *Pineus strobi*
Pinnaspis aspidistrae (Signoret) .. .. fern scale
  Pionea forficalis (L.), see *Evergestis forficalis*
Piophila casei (L.) .. .. (fly) larva = cheese skipper
    cheese maggot
Pissodes castaneus (Degeer) .. .. small banded pine weevil
  Pissodes notatus Fabricius, not *notatus* (Bonsdorff, 1785), see *Pissodes castaneus*
Pissodes pini (L.) .. .. banded pine weevil
Pissodes validirostris (Sahlberg) .. .. pine cone weevil
Pityogenes bidentatus (Herbst) .. .. two-toothed pine beetle
Pityogenes chalcographus (L.) .. .. pine bark beetle
  see also *Hylurgops palliatus*
Pityophthorus pubescens (Marsham) .. pine-twig bark beetle
Plagionotus arcuatus (L.) .. .. .. yellow-bowed longhorn beetle
Planococcus citri (Risso) .. .. .. citrus mealybug
†*Platycheirus manicatus* (Meigen) .. .. potato aphid hover fly
  Platyedra malvella (Hübner), see *Pexicopia malvella*
Platynota idaeusalis (Walker) .. .. tufted apple-bud moth
Platyparea poeciloptera (Schrank) .. asparagus fly
PLATYPODIDAE .. .. ambrosia beetles
  (in particular *Platypus* spp.) larvae = pin-hole borers
  see also *Xyleborus* spp. shot-hole borers
Platypus cylindrus (Fabricius) .. .. (beetle) larva = oak pin-hole borer
  Plectocryptus arrogans (Gravenhorst), see *Polytribax arrogans*
Plesiocoris rugicollis Fallén .. .. apple capsid (bug)
Plinthus caliginosus (Fabricius) .. .. hop root weevil
Plodia interpunctella (Hübner) .. .. Indian meal moth
  Plusia chalcites (Esper), see *Chrysodeixis chalcites*
  Plusia gamma (L.), see *Autographa gamma*
  Plusia ni (Hübner), see *Trichoplusia ni*
  Plusia orichalcea (Fabricius), see *Diachrysia orichalcea*

SCIENTIFIC NAMES—Arthropods

Scientific Name	Common Name
PLUSIINAE	plusia moths
	larvae = *semi-loopers*
Plutella maculipennis (Curtis), see *Plutella xylostella*	
Plutella xylostella (L.)	diamond-back moth
	cabbage web moth
Pnyxia scabiei (Hopkins)	potato scab gnat
Poecilocampa populi (L.)	December moth
Pogonocherus hispidulus (Piller & Mitterpacher)	apple-wood longhorn beetle
Pogonognathellus, subgenus of *Tomocerus*	
Pollenia spp.	cluster flies
	swarming flies
see also *Eudasyphora* spp., *Hydrotaea* spp., *Ophyra* spp. and *Thaumatomyia* spp.	
Pollenia rudis (Fabricius)	cluster fly
Polychrysia moneta (Fabricius)	delphinium moth
	golden plusia moth
Polydesmus angustus Latzel	flat millepede
see also *Brachydesmus superus*	flat-backed millepede
Polydrusus pilosus Gredler	spruce needle weevil
Polydrusus tereticollis: not *tereticollis* (Degeer, 1775), although misidentified as such in Bonsdorff, 1785, and in several later published works, see *Polydrusus undatus*	
Polydrusus undatus (Fabricius)	black-waved weevil
Polygraphus poligraphus (L.)	small spruce bark beetle
Polyphagotarsonemus latus (Banks)	broad mite
Polyplax spinulosa (Burmeister)	rat louse
†*Polytribax arrogans* (Gravenhorst)	pine looper parasite (wasp)
see also *Cratichneumon viator*, *Dusona oxyacanthae* and *Heteropelma calcator*	
Pomaphis, subgenus of *Dysaphis*	
Pontania spp.	willow leaf gall sawflies
Pontania harrisoni Benson, see *Pontania viminalis*	
Pontania pedunculi (Hartig) } *Pontania viminalis* (L.) }	willow pea-gall sawflies
Pontania proxima (Lepeletier)	willow bean-gall sawfly
Popillia japonica Newman	Japanese beetle
Porcellio scaber Latreille	garden woodlouse
Porthesia chrysorrhoea (L.), see *Euproctis chrysorrhoea*	
Porthetria dispar (L.), see *Lymantria dispar*	
Prays curtisella (Donovan), see *Prays fraxinella*	
Prays fraxinella (Bjerkander)	ash bud moth
Prionus coriarius (L.)	oak longhorn beetle
	tanner beetle
see also *Rhagium mordax*	
Priophorus brullei Dahlbom, see *Priophorus morio*	
Priophorus morio (Lepeletier)	small raspberry sawfly
Priophorus padi: not *padi* (Linnaeus, 1761), although misidentified as such in several later published works, see *Priophorus pallipes*	
Priophorus pallipes (Lepeletier)	plum leaf sawfly
Priophorus tener (Hartig): attribution to author incorrect, see next entry	
Priophorus tener (Zaddach), see *Priophorus morio*	
Priophorus varipes (Lepeletier), see *Priophorus pallipes*	
Priophorus viminalis (Fallén), see *Trichiocampus viminalis*	
Pristiphora abietina (Christ)	gregarious spruce sawfly
Pristiphora alnivora (Hartig)	columbine sawfly
Pristiphora alpestris (Konow)	birch sawfly
Pristiphora ambigua (Fallén)	spruce tip sawfly
Pristiphora conjugata (Dahlbom)	poplar sawfly
Pristiphora erichsonii (Hartig)	large larch sawfly
Pristiphora furvescens (Cameron), see *Pristiphora ambigua*	
Pristiphora laricis (Hartig)	small larch sawfly

SCIENTIFIC NAMES—Arthropods

Scientific Name	Common Name
Pristiphora pallipes Lepeletier	small gooseberry sawfly
Pristiphora pini (Retzius), see *Pristiphora abietina*	
Pristiphora pseudocoactula (Lindqvist) ⎫	
Pristiphora quercus (Hartig) ⎬ ..	birch sawflies
Pristiphora testacea (Jurine) ⎭	
see also *Pristiphora alpestris*	
Pristiphora subartica (Forsslund)	Scandinavian spruce sawfly
Pristiphora subbifida (Thomson)	sycamore sawfly
Pristiphora wesmaeli (Tischbein)	larch sawfly
see also *Pachynematus imperfectus*	
Prociphilus pini (Burmeister), see *Stagona pini*	
Procus literosa (Haworth), see *Mesoligia literosa*	
Procus strigilis (Clerck): attribution to author incorrect, see *Oligia strigilis* (Linnaeus)	
Prodenia littoralis (Boisduval), see *Spodoptera littoralis*	
Prodenia litura: not *litura* (Fabricius, 1775), although misidentified as such in several later published works, see *Spodoptera littoralis*	
Proisotoma minuta (Tullberg)	minute grey springtail
Propylea 14-punctata (L.), see *Propylea quattuordecimpunctata*	
†*Propylea quattuordecimpunctata* (L.) ..	fourteen-spot ladybird (beetle)
Protapion, subgenus of *Apion*	
Protemphytus carpini (Hartig)	geranium sawfly
Protemphytus pallipes (Spinola)	viola sawfly
Protoemphytus: misspelling, see *Protemphytus*	
Protolachnus bluncki (Börner), see *Eulachnus rileyi*	
†*Psallus ambiguus* (Fallén)	red apple capsid (bug)
Pseudaulacaspis pentagona (Targioni-Tozzetti)	white peach scale
Pseudochermes fraxini (Kaltenbach) ..	ash scale
PSEUDOCOCCIDAE	mealybugs
Pseudococcus adonidum: not *adonidum* (L.), although misidentified as such in several later published works, see *Pseudococcus longispinus*	
Pseudococcus affinis (Maskell)	glasshouse mealybug
	vine mealybug (see also *Pseudococcus maritimus*)
Pseudococcus citri (Risso), see *Planococcus citri*	
Pseudococcus calceolariae (Maskell) ..	citrophilus mealybug
	currant mealybug
Pseudococcus fragilis Brain, see *Pseudococcus calceolariae*	
Pseudococcus gahani Green, see *Pseudococcus calceolariae*	
Pseudococcus longispinus (Targioni-Tozzetti)	long-tailed mealybug
Pseudococcus mamillariae: not *mamillariae* (Bouché, 1844), although misidentified as such in Green, 1930, and in several later published works, see *Spilococcus cactearum*	
Pseudococcus maritimus (Ehrhorn)	vine mealybug
note: the name *maritimus* was wrongly used in Green, 1920–31, and in several later published works, see *Pseudococcus affinis*	
Pseudococcus obscurus Essig, see *Pseudococcus affinis*	
Pseudophonus, subgenus of *Harpalus*	
Pseudopityophthorus spp.	American oak bark beetles
PSEUDOSCORPIONES	false scorpions
	pseudoscorpions
Psila nigricornis Meigen	*larva* = chrysanthemum stool miner
	lettuce root fly
Psila rosae (Fabricius)	carrot fly
PSOCOPTERA	psocids
	booklice
	barklice
	dustlice
Psorergates spp.	itch mites
see also *Sarcoptes scabiei*	follicle mites (see also DEMODICIDAE)

SCIENTIFIC NAMES—Arthropods

Scientific Name   **Common Name**

Psoroptes bovis (Hartwig) ⎫
Psoroptes cuniculi (Delafond) ⎬ see *Psoroptes equi*
Psoroptes ovis (Hering) ⎭
Psoroptes equi (Raspail) .. .. .. psoroptic mange mite
                                           ox psoroptic mange mite
                                           rabbit ear canker mite
                                           rabbit ear mite
                                           sheep scab mite
PSOROPTIDAE .. .. scab mites
   see also EPIDERMOPTIDAE and PYROGLYPHIDAE
PSYCHODIDAE .. .. moth flies
                                             owl midges
                                             sewage farm flies
                                             waltzing midges
Psylla alni (L.) .. .. alder sucker (psyllid)
Pyslla buxi (L.) .. .. box sucker (psyllid)
Psylla mali (Schmidberger) .. .. .. apple sucker (psyllid)
Psylla pyricola Förster .. .. .. pear sucker
                                             pear psylla
                                             pear psyllid
   Psylla simulans Förster, see *Psylla pyricola*
Psylla visci Curtis .. .. mistletoe sucker (psyllid)
PSYLLIDAE .. .. psyllids
                                             jumping plant lice
                                             suckers
Psylliodes affinis (Paykull) .. .. .. potato flea beetle
Psylliodes attenuata (Koch) .. .. .. hop flea beetle
Psylliodes chrysocephala (L.) .. .. cabbage stem flea beetle
Psylliodes hyoscyami (L.) .. .. .. henbane flea beetle
Psyllipsocus ramburii Sélys-Longchamps .. pale booklouse
                                             vinegar barrel psocid
   Psyllobora vigintiduopunctata (L.), see *Thea vigintiduopunctata*
Psyllopsis fraxini (L.) .. .. .. ash leaf gall sucker (psyllid)
Psyllopsis fraxinicola (Förster) .. .. ash leaf sucker (psyllid)
  Pterochlorus rosae (Cholodkovsky), see *Maculolachnus submacula*
  Pterochlorus salicis (Sulzer), see *Tuberolachnus salignus*
  Pterochlorus saligna (Gmelin), see *Tuberolachnus salignus*
Pterocomma salicis (L.) .. .. .. black willow aphid
†*Pteromalus puparum* (L.) .. .. .. white butterfly parastic wasp
  Pteronidea leucotrocha (Hartig), see *Nematus leucotrochus*
  Pteronidea melanaspis (Hartig), see *Nematus melanaspis*
  Pteronidea pavida (Lepeletier), see *Nematus pavidus*
  Pteronidea ribesii (Scopoli), see *Nematus ribesii*
  Pteronidea salicis (L.), see *Nematus salicis*
  Pteronidea spiraeae (Zaddach & Brischke), see *Nematus spiraeae*
Pterostichus spp. .. .. cloakers
Pterostichus cupreus (L.) .. .. .. rain beetle
                                             strawberry ground beetle (see also
                                                   Pterostichus madidus and *Pterostichus*
                                                   melanarius)
Pterostichus madidus (Fabricius) ⎫
Pterostichus melanarius (Illiger) ⎬ .. strawberry ground beetles
  see also *Pterostichus cupreus* ⎭
   Pterostichus vulgaris: not *vulgaris* (Linnaeus, 1758), although misidentified as such in several
     later published works, see *Pterostichus melanarius*
   Pthirus inguinalis Leach, see *Pthirus pubis*
   Pthirus inquinalis Leach: misspelling, see entry above
Pthirus pubis (L.) .. .. crab louse
                                             pubic louse

SCIENTIFIC NAMES—Arthropods

Scientific Name	Common Name
Ptilinus pectinicornis (L.)	post boring beetle
PTINIDAE	spider beetles
Ptinus brunneus Duftschmid, see *Ptinus clavipes*	
Ptinus clavipes Panzer	brown spider beetle
Ptinus fur (L.)	white-marked spider beetle
Ptinus hirtellus Sturm, see *Ptinus clavipes*	
Ptinus tectus Boieldieu	Australian spider beetle
Ptinus testaceus Olivier, see *Ptinus clavipes*	
Ptinus villiger Reitter	hairy spider beetle
Ptycholoma lecheana (L.)	Leche's twist moth
Pulex irritans L.	human flea
Pulvinaria floccifera (Westwood), see *Chloropulvinaria floccifera*	
Pulvinaria regalis Canard	horse-chestnut scale
Pulvinaria ribesiae Signoret	woolly currant scale
Pulvinaria vitis (L.)	woolly vine scale
Pycnoscelus surinamensis (L.)	Surinam cockroach
Pyemotes tritici (La Grèze-Fossat & Montagné)	straw itch mite / grain itch mite
Pyemotes ventricosus (Newport), see *Pyemotes tritici*	
PYGMEPHORIDAE (in particular *Pygmephorus* spp.)	red pepper mites
Pyralis costalis (Fabricius), see *Hypsopygia costalis*	
Pyralis farinalis (L.)	meal moth / meal snout moth
Pyrausta nubilalis (Hübner), see *Ostrinia nubilalis*	
Pyrellia lasiophthalma: not *lasiophthalma* (Macquart, 1835), although misidentified as such in several later published works, see *Eudasyphora cyanella*	
*Pyroderces rileyi (Walsingham)	(moth) *larva* = pink scavenger
PYROGLYPHIDAE	dust mites / scab mites (see also PSOROPTIDAE and EPIDERMOPTIDAE)
Pyrrhalta viburni (Paykull)	viburnum beetle
*Pyrrhidium sanguineum (L.)	scarlet-coated longhorn beetle
Quadraspidiotus ostreaeformis (Curtis) } *Quadraspidiotus pyri* (Lichtenstein) }	oystershell scales (*Quadraspidiotis ostreaeformis* also known as oyster scale, *Quadraspidiotis pyri* also known as pear scale)
Quadraspidiotus perniciosus (Comstock), see *Comstockaspis perniciosa*	
Radfordia affinis (Poppe)	mouse fur mite / mouse myobiid mite
Radfordia ensifera (Poppe)	rat fur mite / rat myobiid mite
Raphidia notata Fabricius	common snake-fly
RAPHIDIOPTERA	snake-flies
Reduvius personatus (L.)	masked hunter bug / fly bug
Resseliella crataegi (Barnes)	hawthorn stem midge
Resseliella oculiperda (Rübsaamen)	(fly) *larva* = red bud borer
Resseliella theobaldi (Barnes)	raspberry cane midge
Rhabdophaga clausilia (Bremi)	willow leaf-folding midge / osier leaf-folding midge (see also *Rhabdophaga marginemtorquens*)

SCIENTIFIC NAMES—Arthropods

Scientific Name	Common Name
Rhabdophaga heterobia (Löw)	willow button-top midge
Rhabdophaga marginemtorquens (Bremi)	osier leaf-folding midge
see also *Rhabdophaga clausilia*	
Rhabdophaga justini Barnes	
Rhabdophaga purpureaperda Barnes ⎫	willow shot-hole midges
Rhabdophaga triandraperda Barnes ⎭	
see also *Helicomyia saliciperda*	
Rhabdophaga rosaria (Loew)	willow rosette-gall midge
	terminal rosette-gall midge
Rhabdophaga salicis (Schrank)	willow stem gall midge
Rhabdophaga terminalis (Loew)	willow terminal leaf midge
Rhadinoceraea micans (Klug)	iris sawfly
RHAGIONIDAE	snipe flies
(in particular *Atherix* spp. and *Symphoromyia* spp.)	
Rhagium mordax (Degeer)	oak longhorn beetle
see also *Prionus coriarius*	
Rhagoletis alternata (Fallén)	rose-hip fly
*Rhagoletis cerasi (L.)	European cherry fruit fly
*Rhagoletis cingulata (Loew)	North American cherry fruit fly
*Rhagoletis fausta (Osten Sacken)	black cherry fruit fly
*Rhagoletis pomonella (Walsh)	apple fruit fly
	larva = apple maggot
*Raphuma annularis (Fabricius)	yellow bamboo longhorn beetle
Rhipicephalus sanguineus (Latreille)	kennel tick
	brown dog tick
	tropical brown dog tick
Rhizarcha spp., see *Dacnusa*	
Rhizoecus spp.	root mealybugs
Rhizoglyphus callae Oudemans ⎫	bulb mites
Rhizoglyphus robini Claparède ⎭	

Rhizoglyphus echinopus: not *echinopus* (Fumouze & Robin, 1868), although at least two different species have been misidentified as such in several later published works, see *Rhizoglyphus callae* and *Rhizoglyphus robini*

Rhizopertha, see *Rhyzopertha*
Rhodites nervosus (Curtis), see *Diplolepis nervosa*
Rhodites rosae (L.), see *Diplolepis rosae*
Rhodophaea, subgenus of *Eurhodope*

Rhopalomyia chrysanthemi (Ahlberg)	chrysanthemum gall midge
	chrysanthemum midge
Rhopalosiphoninus latysiphon (Davidson)	bulb and potato aphid
Rhopalosiphoninus ribesinus (van der Goot)	currant stem aphid
Rhopalosiphoninus staphyleae (Koch)	
spp. *tulipaellus* (Theobald)	mangold aphid
Rhopalosiphum crataegellum (Theobald), see *Rhopalosiphum insertum*	
Rhopalosiphum insertum (Walker)	apple—grass aphid
	oat—apple aphid
Rhopalosiphum maidis (Fitch)	cereal leaf aphid
	corn leaf aphid
Rhopalosiphum nymphaeae (L.)	water-lily aphid
Rhopalosiphum padi (L.)	bird-cherry aphid
	bird-cherry—oat aphid
Rhopobota naevana (Hübner)	holly tortrix (moth)
	marbled single-dot bell moth
	larva = holly leaf tier
	black-headed fireworm
Rhopobota unipunctana (Haworth), see *Rhopobota naevana*	
Rhyacionia buoliana (Denis & Schiffermüller)	pine shoot moth
Rhyacionia duplana (Hübner)	Elgin shoot moth
Rhynchaenus spp.	leaf mining weevils

SCIENTIFIC NAMES—Arthropods

Scientific Name	Common Name
Rhynchaenus fagi (L.)	beech leaf mining weevil
	larva = beech leaf miner
Rhynchaenus quercus (L.)	oak leaf mining weevil
Rhynchites spp.	rhynchites
	twig cutting weevils
Rhynchites aequatus (L.)	apple fruit rhynchites (weevil)

Rhynchites betuleti (Fabricius), see *Byctiscus betulae*
Rhynchites betuli (Fabricius): misspelling of *betuleti* (Fabricius), see *Byctiscus betulae*

Rhynchites caeruleus (Degeer)	apple twig cutter (weevil)
Rhynchites germanicus Herbst	strawberry rhynchites (weevil)
RHYNCHOPHTHIRINA	biting lice

see also AMBLYCERA and ISCHNOCERA

†*Rhyssa persuasoria* (L.)	wood wasp parasite (wasp)
Rhyzopertha dominica (Fabricius)	(beetle) *larva* = lesser grain borer
Ribautiana debilis (Douglas) *Ribautiana tenerrima* (Herrich-Schäffer)	fruit tree leafhoppers

see also species of *Alnetoidia*, *Edwardsiana*, *Typhlocyba* and *Zygina*

Riccardoella limacum (Schrank)	slug mite

Romaleum rufulum Haldeman, see *Enaphalodes rufulum*

Sacchiphantes, subgenus of *Adelges*
  Sacchiphantes abietis (L.), see *Adelges abietis*
  Sacchiphantes viridis (Ratzburg), see *Adelges viridis*

Saissetia coffeae (Walker)	hemispherical scale

  Saissetia hemisphaerica (Targioni-Tozzetti), see *Saissetia coffeae*

Saissetia oleae (Olivier)	Mediterranean black scale
SALDIDAE	jumping bugs, shore bugs
SALTATORIA	grasshoppers, crickets and bush-crickets
Saperda carcharias (L.)	large poplar longhorn beetle
	larva = poplar and willow borer
Saperda populnea (L.)	small poplar longhorn beetle
	larva = small poplar borer

Sappaphis aucupariae (Buckton), see *Dysaphis aucupariae*
Sappaphis crataegi (Kaltenbach), see *Dysaphis crataegi*
Sappaphis devecta (Walker), see *Dysaphis devecta*
Sappaphis mali (Ferrari), see *Dysaphis plantaginea*
Sappaphis petroselini (Börner): *petroselini* (Börner) is a species synonym, but the name is retained for a subspecies, see *Dysaphis apiifolia*
Sappaphis plantaginea (Passerini), see *Dysaphis plantaginea*
Sappaphis pyri (Boyer de Fonscolombe), see *Dysaphis pyri*
Sappaphis sorbi (Kaltenbach), see *Dysaphis sorbi*
Sappaphis tulipae (Boyer de Fonscolombe), see *Dysaphis tulipae*

SARCOPHAGIDAE	flesh flies

(in particular *Sarcophaga* spp.)
see also CALLIPHORIDAE
Sarcoptes canis Gerlach, see *Sarcoptes scabiei*
Sarcoptes caprae Fürstenberg, see *Sarcoptes scabiei*
Sarcoptes equi Gerlach, see *Sarcoptes scabiei*
Sarcoptes ovis Mégnin, see *Sarcoptes scabiei*
Sarcoptes scabei: misspelling, see *Sarcoptes scabiei*
Sarcoptes scabiei (Degeer): attribution to author incorrect, see *Sarcoptes scabiei* (L.)

SCIENTIFIC NAMES—Arthropods

Scientific Name	Common Name
Sarcoptes scabiei (L.)	itch mite
see also *Psorergates* spp.	dog itch mite
	horse itch mite
	pig itch mite
	sarcoptic mange mite (of goat)
	scabies mite
	sheep itch mite

Sarcoptes suis Gerlach, see *Sarcoptes scabiei*
Sathrobrota rileyi (Walsingham), see *Pyroderces rileyi*

†*Scaeva pyrastri* (L.)	cabbage aphid hover fly
†*Scambus pomorum* (Ratzeburg)	apple blossom weevil parasite (wasp)
SCARABAEIDAE	chafers, dung beetles, etc.
Scatella tenuicosta Collin	glasshouse wing-spot fly
Scathophaga stercoraria (L.)	yellow dung fly
	dung fly

Scatophaga: misspelling, see *Scathophaga*

SCENOPINIDAE	window flies
(in particular *Scenopinus* spp.)	
Schistocerca gregaria (Forskål)	desert locust
Schizolachnus pineti (Fabricius)	grey pine needle aphid

Schizoneura, subgenus of *Eriosoma*
Schizoneura ulmi (L.), see *Eriosoma ulmi*

Sciaphilus asperatus (Bonsdorff)	scaly strawberry weevil
	strawberry root weevil (see also species of *Otiorhynchus*)

Sciaphilus muricatus (Fabricius), see *Sciaphilus asperatus*
Sciara: not *Sciara* Meigen, although misidentified as such in several later published works, see *Bradysia*

SCIOMYZIDAE	slug- and snail-killing flies
Scirtothrips longipennis (Bagnall)	begonia thrips
	long-winged thrips
Scolopostethus pictus (Schilling)	corn-stack bug
	hay-stack bug
SCOLYTIDAE	bark beetles
(in particular *Scolytus* spp.)	

Scolytus destructor Olivier, see *Scolytus scolytus*

Scolytus intricatus (Ratzeburg)	oak bark beetle
Scolytus mali (Bechstein)	large fruit bark beetle
Scolytus multistriatus (Marsham)	small elm bark beetle
Scolytus pruni (Ratzeburg), see *Scolytus mali*	
Scolytus ratzeburgi Janson	birch bark beetle
Scolytus rugulosus (Müller)	fruit bark beetle

Scolytus rugulosus Ratzeburg: attribution to author incorrect, see *Scolytus rugulosus* (Müller)

Scolytus scolytus (Fabricius)	large elm bark beetle
Scrobipalpa artemisiella (Treitschke) ..	thyme moth
SCUTELLERIDAE	shield bugs
see also ACANTHOSOMIDAE, CYDNIDAE and PENTATOMIDAE	
Scutigerella immaculata (Newport)	glasshouse symphylid
	glasshouse 'centipede'
	garden symphylan
Scythropia crataegella (L.)	hawthorn moth
	larva = hawthorn webber
Semanotus undatus (L.)	European pine longhorn beetle

Semiadalis: misspelling, see *Semidalis*
Semiaphis atriplicis (L.), see *Hayhurstia atriplicis*

†*Semidalis aleyrodiformis* (Stephens) ..	powdery lacewing
see also *Coniopteryx* spp. and *Conwentzia* spp.	
Semiothisa liturata (Clerck)	tawny-barred angle moth

SCIENTIFIC NAMES—Arthropods

Scientific Name	Common Name
SEPSIDAE	wing wagger flies
(in particular *Sepsis* spp.)	
Serica brunnea (L.)	brown chafer (beetle)
Serropalpus barbatus (Schaller)	(beetle) *larva* = dry wood borer
Sesia apiformis (Clerck)	hornet moth
	hornet clearwing
Sesia bembeciformis (Hübner)	lunar hornet moth
Siclobola, subgenus of *Clepsis*	
SILPHIDAE	carrion beetles
Simaethis pariana (Clerck), see *Choreutis pariana*	
SIMULIIDAE	black flies
Simulium reptans (L.)	common black fly
Sinodendron cylindricum (L.)	spiked pox beetle
Sinoxylon anale Lesne	common auger beetle
Sinoxylon conigerum Gerstaecker	conifer auger beetle
Siphona irritans (L.), see *Haematobia irritans*	
Siphona stimulans (Meigen), see *Haematobosca stimulans*	
SIPHONAPTERA	fleas
Siphoninus immaculatus (Heeger)	ivy whitefly
Siphoninus phillyreae (Haliday)	phillyrea whitefly
SIPHUNCULATA, see ANOPLURA	
Sirex cyaneus Fabricius ⎱	
Sirex juvencus (L.) ⎬	steel-blue wood wasps
Sirex noctilio Fabricius ⎰	
Sirex gigas (L.), see *Urocerus gigas*	
Siteroptes graminum (Reuter)	grass and cereal mite
	silver top (of cereals and grasses)
Sitobion avenae (Fabricius)	grain aphid
Sitobion fragariae (Walker)	blackberry—cereal aphid
	blackberry aphid
Sitobion luteum (Buckton)	yellow orchid aphid
Sitodiplosis dactylidis Barnes	cocksfoot midge
see also *Dasineura dactylidis* and species of *Contarinia*	
Sitodiplosis mosellana (Géhin)	orange wheat blossom midge
	wheat blossom midge (see also *Contarinia tritici*)
Sitodrepa panicea (L.), see *Stegobium paniceum*	
Sitona spp.	bean weevils, clover weevils, pea weevils
Sitona crinitus (Herbst, 1795), not *crinitus* (Gmelin in Linnaeus, 1789), see *Sitona macularius*	
Sitona flavescens (Marsham, 1802), not *flavescens* (Fabricius, 1787), see *Sitona lepidus*	
Sitona hispidulus (Fabricius) ⎱	common clover weevils
Sitona sulcifrons (Thunberg) ⎰	
Sitona lineatus (L.)	pea and bean weevil
Sitona macularius (Marsham)	broom and clover weevil
Sitona humeralis Stephens ⎱	
Sitona lepidus Gyllenhal ⎬	clover weevils
Sitona puncticollis Stephens ⎰	
Sitophilus granarius (L.)	grain weevil
Sitophilus oryzae (L.)	rice weevil
	lesser rice weevil
Sitophilus zeamais Motschulsky	maize weevil
	greater rice weevil
Sitotroga cerealella (Olivier)	Angoumois grain moth
Smerinthus ocellata (L.)	eyed hawk moth
Sminthurus viridis (L.)	lucerne-flea (springtail)
Smynthurodes betae Westwood	bean root aphid
Smynthurus, see *Sminthurus*	
Solenopotes capillatus Enderlein	blue cattle louse
	small blue cattle louse
	tubercle-bearing louse

SCIENTIFIC NAMES—Arthropods

Scientific Name **Common Name**

Spaelotis obscura (Brahm), see *Spaelotis ravida*
Spaelotis ravida (Denis & Schiffermüller) .. stout dart moth
Spectrobates ceratoniae (Zeller), see *Ectomyelois ceratoniae*
SPHAEROCERIDAE .. .. .. lesser dung flies
Sphecia bembeciformis (Hübner), see *Sesia bembeciformis*
SPHINGIDAE .. .. hawk moths
Spilococcus cactearum McKenzie .. .. cactus mealybug
               cactus root mealybug
 Spilographa alternata (Fallén), see *Rhagoletis alternata*
 Spilographa zoe (Meigen), see *Trypeta zoe*
Spilonota ocellana (Denis & Schiffermüller)   bud moth
               eye-spotted bud moth
               larva = brown apple budworm
 Spilonota uddmanniana (L.), see *Epiblema uddmanniana*
Spilopsyllus cuniculi (Dale) .. .. .. European rabbit flea
               rabbit flea
Spilosoma lubricipede (L.) .. .. .. white ermine moth
Spilosoma lutea (Hufnagel) .. .. .. buff ermine moth
Spiniphora bergenstammii (Mik) .. .. milkbottle scuttle fly
Spodoptera exigua (Hübner) .. .. small mottled willow moth
               larva = beet armyworm
               lesser armyworm
**Spodoptera littoralis* (Boisduval) .. .. (moth) *larva* = African cotton leafworm
               Egyptian cotton leafworm
               Mediterranean climbing
                cutworm
               adult = Mediterranean brocade
**Spodoptera litura* (Fabricius) .. .. (moth) *larva* = Asian cotton leafworm
               Asian and Pacific
                cutworm
 note: the name *litura* has been wrongly used in several published works to describe *Spodoptera littoralis* (Boisduval)
Spuleria atra (Haworth) .. .. .. pith moth
               apple pith moth
Stagona pini (Burmeister) .. .. .. pine root aphid
STAPHYLINIDAE .. .. rove beetles
Staphylinus olens Müller .. .. .. devil's coach-horse beetle
 Steganoptycha diniana (Guenée), see *Zeiraphera diniana*
Stegobium paniceum (L.) .. .. .. biscuit beetle
               bread beetle
               drug store beetle
 Steneotarsonemus fragariae (Zimmerman), see *Phytonemus pallidus fragariae*
Steneotarsonemus laticeps (Halbert) .. bulb scale mite
 Steneotarsonemus pallidus (Banks), see *Phytonemus pallidus*
Steneotarsonemus spirifex (Marchal) .. oat spiral mite
 Stenodiplosis geniculati Reuter, see *Contarinia geniculati*
Stenothrips graminum Uzel .. .. .. oats thrips
Stephanitis rhododendri Horváth .. .. rhododendron bug
 Stephanoderes hampei (Ferrari), see *Hypothenemus hampei*
Sternostoma tracheacolum Lawrence .. canary lung mite
†*Stethorus punctillum* Weise .. .. minute black ladybird (beetle)
Stigmella anomalella (Goeze) .. .. (moth) *larva* = rose leaf miner
Stigmella malella (Stainton) .. .. .. apple pygmy moth
               larva = crab apple leaf miner
 Stigmella rosella (Schrank), see *Stigmella anomalella*
Stilbus spp. .. .. shining smut beetles
 Stilpnotia salicis (L.), see *Leucoma salicis*
Stomoxys calcitrans (L.) .. .. .. stable fly
               biting house fly

SCIENTIFIC NAMES — Arthropods

Scientific Name	Common Name
STRATIOMYIDAE	soldier flies
Stromatium fulvum (Villers)	Arabian red longhorn beetle

Strophosomus coryli (Fabricius), see *Strophosomus melanogrammus*
Strophosomus lateralis (Paykull, 1792), not *lateralis* (Panzer, 1789), see *Strophosomus sus*

Strophosomus melanogrammus (Förster)	nut leaf weevil
Strophosomus sus Stephens	heather weevil

Subcoccinella 24-punctata (L.), see *Subcoccinella vigintiquattuorpunctata*

Subcoccinella vigintiquattuorpunctata (L.)	twentyfour-spot ladybird (beetle)
Suidasia nesbitti Hughes	scaly grain mite wheat-pollard itch mite
Supella longipalpa (Fabricius)	brown-banded cockroach

Supella supellectilium (Serville), see *Supella longipalpa*

Sylvicola spp.	{ sewage filter-bed flies { window gnats
Sylvicola fenestralis (Scopoli)	common window gnat
Symphoromyia spp.	snipe flies

see also *Atherix* spp.

SYMPHYLA	symphylids
Symphylella spp.	open ground symphylids
Synanthedon myopaeformis (Borkhausen)	apple clearwing moth red-belted clearwing

Synanthedon salmachus (L.), see *Synanthedon tipuliformis*

Synanthedon tipuliformis (Clerck)	currant clearwing moth *larva* = currant borer
Syndemis musculana (Hübner)	afternoon twist moth

Syndiplosis petioli (Kieffer), see *Contarinia petioli*

Syringophilus bipectinatus Heller	quill mite (of poultry)
SYRPHIDAE	hover flies
(in particular *Syrphus* spp.)	*larvae* = greenfly predators
†*Syrphus ribesii* (L.)	currant hover fly

TABANIDAE	clegs, gadflies, horseflies breeze flies dun flies stouts
Tabanus bromius L.	small horse fly
Tabanus bovinus L. } *Tabanus sudeticus* Zeller }	large horse flies
TACHINIDAE	parasitic flies
Tachycines asynamorus Adelung	glasshouse camel-cricket

Tachypodoiulus albipes (Koch), see *Tachypodoiulus niger*

Tachypodoiulus niger (Leach)	white-legged black millepede black millepede (see also *Cylindroiulus londinensis*) snake millepede (see also *Archiboreoiulus pallidus*, *Boreoiulus tenuis* and *Blaniulus guttulatus*)

Taeniocampa incerta (Hufnagel), see *Orthosia incerta*
Taeniocampa instabilis (Denis & Schiffermüller), see *Orthosia incerta*
Taeniothrips atratus (Haliday), see *Thrips atratus*
Taeniothrips britteni (Bagnall), see *Thrips atratus*
Taeniothrips gladioli Moulton & Steinweden, see *Thrips simplex*

Taeniothrips inconsequens (Uzel)	pear thrips fruit tree thrips
Taeniothrips laricivorus Kratochvil & Farský	larch thrips

Taeniothrips pyri (Daniel), see *Taeniothrips inconsequens*
Taeniothrips simplex (Morison), see *Thrips simplex*

SCIENTIFIC NAMES—Arthropods

Scientific Name	Common Name
Tanymecus palliatus (Fabricius)	beet leaf weevil
TARSONEMIDAE	tarsonemid mites

    Tarsonemus approximatus Banks, see *Steneotarsonemus laticeps*
    Tarsonemus fragariae Zimmerman, see *Phytonemus pallidus fragariae*
    Tarsonemus laticeps Halbert, see *Stenotarsonemus laticeps*
    Tarsonemus latus Banks, see *Polyphagotarsonemus latus*

Tarsonemus myceliophagus Hussey	mushroom mite
see also *Tyrophagus* spp.	

    Tarsonemus pallidus Banks, see *Phytonemus pallidus*
    Tarsonemus spirifex Marchal, see *Steneotarsonemus spirifex*

Taxomyia taxi (Inchbald)	yew gall midge

    Taxonus glabratus (Fallén), see *Ametastegia glabrata*

Tegenaria spp.	house spiders
Teichomyza fusca Macquart	urinal fly
Tenebrio molitor L.	yellow mealworm beetle
	larva = mealworm
Tenebrio obscurus Fabricus	dark mealworm beetle
	larva = dark mealworm
TENEBRIONIDAE	tenebrionid beetles
Tenebroides mauritanicus (L.)	cadelle beetle
TEPHRITIDAE	large fruit flies
	gall flies
Tetraneura ulmi (L.)	elm—grass-root aphid
	elm leaf gall aphid
TETRANYCHIDAE	spider mites

    Tetranychus bimaculatus Harvey, see *Tetranychus urticae*

Tetranychus cinnabarinus (Boisduval)	carmine spider mite

    Tetranychus telarius: description of Linnaeus, 1758, is unresolved; in later published works the name has been applied to *Tetranychus urticae*
    Tetranychus tiliarium (Hermann), see *Eotetranychus tiliarium*

Tetranychus urticae Koch	two-spotted spider mite
	glasshouse red spider mite
	hop red spider mite
	red spider mite
Tetranychus viennensis Zacher	hawthorn spider mite
†*Tetrastichus eriophyes* Taylor	filbert bud mite parasite (wasp)
**Tetropium castaneum* (L.)	chestnut longhorn beetle
**Tetropium cinnamopterum* Kirby	rust-winged longhorn beetle

    Tetropium fuscum: not *fuscum* (Fabricius, 1787), although misidentified as such in several later published works, see *Tetropium gabrieli*

Tetropium gabrieli Weise	larch longhorn beetle
Tetrops praeusta (L.)	little longhorn beetle
Tettigonia viridissima (L.)	great green bush-cricket
TETTIGONIIDAE	bush-crickets
	long-horned grasshoppers
†*Thanasimus formicarius* (L.)	ant beetle
Thaumatomyia spp.	cluster flies
	swarming flies
see also *Eudasyphora* spp., *Hydrotaea* spp., *Ophyra* spp. and *Pollenia* spp.	
Thaumatomyia notata (Meigen)	yellow swarming fly
Thea 22-punctata (L.), see *Thea vigintiduopunctata*	
†*Thea vigintiduopunctata* (L.)	twentytwo-spot ladybird (beetle)
Thecabius affinis (Kaltenbach)	poplar—buttercup aphid
	poplar leaf gall aphid
Thecabius auriculae (Murray)	auricula root aphid
Thecodiplosis brachyntera (Schwägrichen)	pine needle-shortening gall midge

    Theobaldia annulata (Schrank), see *Culiseta annulata* (Schrank)

Thera firmata (Hübner)	red pine carpet moth

SCIENTIFIC NAMES—Arthropods

Scientific Name	Common Name
Thera obeliscata (Hübner)	grey pine carpet moth
THEREVIDAE	stiletto flies
(in particular *Thereva* spp.)	
Thereva nobilitata (Fabricius)	common stiletto fly
Theria primaria (Haworth)	early moth

Theria rupicapraria: not *rupicapraria* (Denis & Schiffermüller, 1775), although misidentified as such in several later published works, see *Theria primaria*

Thermobia domestica (Packard)	firebrat (bristle tail)

Thomasiniana crataegi Barnes, see *Resseliella crataegi*
Thomasiniana oculiperda (Rübsaamen), see *Resseliella oculiperda*
Thomasiniana theobaldi Barnes, see *Resseliella theobaldi*

Thrips angusticeps Uzel	field thrips
	cabbage thrips
	flax thrips (see also *Thrips lini*)

Thrips asemus Williams, see *Thrips angusticeps*

Thrips atratus Haliday	carnation thrips

Thrips debilis Bagnall, see *Thrips tabaci*

Thrips flavus Schrank	yellow flower thrips
	honeysuckle thrips

Thrips frankeniae Bagnall, see *Thrips flavus*

Thrips fuscipennis Haliday } *Thrips major* Uzel }	rose thrips
	(*Thrips major* also known as rubus thrips)

Thrips linarius Uzel, see *Thrips lini*

*Thrips lini Ladureau	flax thrips

see also *Thrips angusticeps*
Thrips menyanthidis Bagnall, see *Thrips fuscipennis*

Thrips nigropilosus Uzel	chrysanthemum thrips
*Thrips palmi Karny	palm thrips

see also *Parthenothrips dracaenae*

Thrips simplex (Morison)	gladiolus thrips
Thrips tabaci Lindeman	onion thrips
	potato thrips
Thylodrias contractus Motschulsky	odd beetle
Thyreophagus entomophagus (Laboulbène)	museum mite
	flour mite (see also *Acarus siro*)
THYSANOPTERA	thrips
	thunderflies
THYSANURA	bristletails

Tinaea, see *Tinea*
Tinea cloacella Haworth, see *Nemapogon cloacella*
Tinea fuscipunctella Haworth, see *Niditinea fuscella*
Tinea granella (L.), see *Nemapogon granella*

Tinea pallescentella Stainton	large pale clothes moth
Tinea pellionella (L.)	case-bearing clothes moth
	case-making clothes moth
Tineola bisselliella (Hummel)	common clothes moth
	larva = webbing clothes maggot
TINGIDAE	lace bugs
Tipula oleracea L. } *Tipula paludosa* Meigen }	common crane flies
	marsh crane fly (*Tipula paludosa*)
	larvae = leatherjackets
TIPULIDAE	crane flies
(in particular *Tipula* spp.)	daddy longlegs
Tomicus minor (Hartig)	lesser pine shoot beetle
Tomicus piniperda (L.)	pine shoot beetle
Tomocerus longicornis Müller	long-horned springtail

SCIENTIFIC NAMES—Arthropods

Scientific Name	Common Name
TORTRICIDAE	tortrices
	tortrix moths
	larvae = leaf rollers
	rose maggots

Tortrix paleana (Hübner), see *Aphelia paleana*
Tortrix postvittana Walker, see *Epiphyas postvittana*
Tortrix pronubana Hübner, see *Cacoecimorpha pronubana*
Tortrix viburnana Denis & Schiffermüller, see *Aphelia viburnana*

Tortrix viridana (L.)	green oak tortrix (moth)
	oak roller moth
	larva = oak leaf roller
Trachelus tabidus (Fabricius)	black grain stem sawfly
Trachyderes hilaris Bates	South American longhorn beetle
Trama troglodytes von Heyden	Jerusalem artichoke tuber aphid
	artichoke tuber aphid

Triaena, subgenus of *Acronicta*

Trialeurodes vaporariorum (Westwood)	glasshouse whitefly
Tribolium spp.	flour beetles
Tribolium castaneum (Herbst)	rust-red flour beetle
Tribolium confusum Jacquelin du Val	confused flour beetle
Tribolium destructor Uyttenboogaart	dark flour beetle
Trichiocampus viminalis (Fallén)	poplar sawfly

see also *Pristiphora conjugata*

Trichiosoma tibiale Stephens	hawthorn sawfly
Trichocera saltator (Harris)	winter gnat
Trichodectes canis (Degeer)	dog biting louse

Trichodectes latus Nitzsch, see *Trichodectes canis*

Trichodectes melis (Fabricius)	badger louse

Trichodectes octopunctatus Denny, see *Trichodectes canis*

Trichophaga tapetzella (L.)	white-tip clothes moth
	tapestry moth
Trichoplusia ni (Hübner)	ni moth
Trigonaspis megaptera (Panzer)	oak leaf kidney-gall cynipid (wasp)
Trigonogenius globulus Solier	globular spider beetle
Trinophyllum cribratum Bates	(beetle) *larva* = Indian oak borer
Trionymus diminutus Leonardi	New Zealand flax mealybug
Trioza alacris Flor	bay sucker (psyllid)
Trioza apicalis Förster	carrot sucker (psyllid)
Trioza remota Förster	oak leaf sucker (psyllid)

Triphaena pronuba (L.), see *Noctua pronuba*

Trisetacus pini (Nalepa)	pine twig-knot mite
	pine gall mite
	pine twig gall mite

Tritoconicera, subgenus of *Conicera*

Trogium pulsatorium (L.)	larger pale booklouse
Trogoderma granarium Everts	khapra beetle
Trogoderma inclusum LeConte	large cabinet beetle

Trogoderma versicolor: not *versicolor* (Creutzer, 1799), although misidentified as such in several later published works, see *Trogoderma inclusum*
Trombicula autumnalis (Shaw), see *Neotrombicula autumnalis*

Trypeta zoe (Meigen)	(fly) *larva* = chrysanthemum blotch miner

TRYPETIDAE, see TEPHRITIDAE
Trypodendron domesticum (L.), see *Xyloterus domesticus*
Trypodendron lineatum (Olivier), see *Xyloterus lineatus*

Tuberculoides annulatus (Hartig)	oak leaf aphid
Tuberolachnus salignus (Gmelin)	large willow aphid
	giant willow aphid

Tubula, subgenus of *Epichoristodes*
Tylos corrigiolatus (L.), see *Micropeza corrigiolata*

SCIENTIFIC NAMES—Arthropods
Scientific Name Common Name

Typhaea stercorea (L.) .. .. .. hairy fungus beetle
  see also *Mycetaea hirta*   hairy cellar beetle
  Typhlocyba avellanae Edwards, see *Edwardsiana avellanae*
  Typhlocyba crataegi Douglas, see *Edwardsiana crataegi*
  Typhlocyba cruenta Herrich-Schäffer, see *Fagocyba cruenta*
  Typhlocyba debilis Douglas, see *Ribautiana debilis*
  Typhlocyba douglasi Edwards, see *Fagocyba cruenta*
  Typhlocyba froggatti Baker, see *Edwardsiana crataegi*
  Typhlocyba hippocastani Edwards, see *Edwardsiana hippocastani*
  Typhlocyba jucunda Herrich-Shäffer, see *Eupterycyba jucunda*
  Typhlocyba prunicola Edwards, see *Edwardsiana prunicola*
Typhlocyba quercus (Fabricius) .. .. fruit tree leafhopper
  see also species of *Alnetoidia, Edwardsiana, Ribautiana* and *Zygina*
  Typhlocyba rosae (L.), see *Edwardsiana rosae*
  Typhlocyba tenerrima Herrich-Schäffer, see *Ribautiana tenerrima*
TYPHLOCYBIDAE, see CICADELLIDAE
  Typhlodromus finlandicus (Oudemans), see *Amblyseius finlandicus*
†*Typhlodromus pyri* Scheuten .. .. fruit tree red spider mite predator (mite)
  see also *Amblyseius finlandicus*
Tyria jacobaeae (L.) .. .. cinnabar moth
  Tyroglyphus lintneri Osborn, see *Tyrophagus putrescentiae*
  Tyroglyphus longior Gervais, see *Tyrophagus longior*
Tyrolichus casei (Oudemans) .. .. cheese mite
  see also *Tyrophagus* spp. (in particular
    Tyrophagus putrescentiae)
Tyrophagus spp. .. .. fungal mites
        cheese mites (in particular *Tyrophagus putrescentiae*, see also *Tyrolichus casei*)
        mushroom mites (see also *Tarsonemus myceliophagus*)
  Tyrophagus casei (Oudemans), see *Tyrolichus casei*
  Tyrophagus castellanii Hirst, see *Tyrophagus putrescentiae*
Tyrophagus longior (Gervais) .. .. grainstack mite
        cucumber mite (see also *Tyrophagus neiswanderi*)
        French 'fly'
        seed mite
        straw mite
Tyrophagus neiswanderi Johnston & Bruce   cucumber mite
  see also *Tyrophagus longior*
Tyrophagus putrescentiae (Schrank) .. mould mite
        cheese mite (see also *Tyrophagus* spp. and *Tyrolichus casei*)
        copra mite
Tyrophagus similis Volgin .. .. .. grassland mite
  Tyrophagus tenuiclavus Zachvatkin, see *Tyrophagus longior*

Urocerus gigas (L.) .. .. giant wood wasp
        greater horntail
  Urophora zoe (Meigen), see *Trypeta zoe*

  Vasates fockeui (Nalepa & Trouessart), see *Aculus fockeui*
Vespa crabro L. .. .. hornet
  Vespa germanica Fabricius, see *Vespula germanica*
  Vespa vulgaris L., see *Vespula vulgaris*

SCIENTIFIC NAMES—Arthropods

Scientific Name   **Common Name**

VESPIDAE .. .. wasps
Vespula austriaca (Panzer) .. .. .. cuckoo wasp
Vespula germanica (Fabricius) .. .. German wasp
    Vespula norwegica (Fabricius), see *Dolichovespula norwegica*
Vespula rufa (L.) .. .. red wasp
    Vespula sylvestris (Scopoli), see *Dolichovespula sylvestris*
Vespula vulgaris (L.) .. .. common wasp
Viminia, subgenus of *Acronicta*
    Virachola antalus (Hopffer), see *Hypokopelates antalus*
    Viteus vitifoliae (Fitch), see *Daktulosphaira vitifoliae*
    Viteus vitifolii(Fitch), see *Daktulosphaira vitifoliae*

Wachtliella ericina (Loew) .. .. .. erica gall midge
Wachtliella persicariae (L.) .. .. .. polygonum gall midge
    Wachtliella rosarum (Hardy), see *Dasineura rosarum*
Werneckiella equi (Denny) .. .. .. horse biting louse
†*Wesmaelius* spp. .. .. brown lacewings
    see also *Hemerobius* spp.

†*Xantholinus* spp. .. .. xantholine rove beetles
Xenopsylla cheopis (Rothschild) .. .. oriental rat flea
Xenylla spp.
Xenylla mucronata Axelson   .. .. mushroom springtails
Xenylla welchi Folsom
    see also species of *Hypogastrura*
Xeris spectrum (L.) .. .. spectrum wood wasp
Xestia c-nigrum(L.) .. .. setaceous Hebrew character moth
                 larva = spotted cutworm
Xestobium rufovillosum (Degeer) .. .. death watch beetle
    Xestobium tessellatum (de Villers), see *Xestobium rufovillosum* (Degeer)
Xestophanes spp. .. .. potentilla gall cynipids (wasps)
Xiphydria camelus (L.) .. .. .. alder wood wasp
Xiphydria prolongata (Fourcroy) .. .. willow wood wasp
                  willow boring sawfly
Xyleborus spp. .. .. ambrosia beetles
    see also PLATYPODIDAE   *larvae* = pin-hole borers
                       shot-hole borers
Xyleborus dispar (Fabricius) .. .. .. (beetle) *larva* = broad-leaved pinhole borer
Xyleborus dryographus (Ratzeburg) .. broad-leaved wood ambrosia beetle
    see also *Xyloterus domesticus*
    Xyleborus dryophagus (Ratzeburg): misspelling, see *Xyleborus dryographus*
Xyleborus saxeseni (Ratzeburg) .. .. fruit-tree wood ambrosia beetle
Xylena exsoleta (L.) .. .. sword-grass moth
Xylena vetusta (Hübner) .. .. .. red sword-grass moth
**Xyleutes ceramica* (Walker) .. .. .. (moth) *larva* = bee-hole borer
*Xylion adustus (Fåhraeus) .. .. .. (beetle) *larva* = Madagascan wood borer
    see also *Xylopsocus capucinus*
**Xylobiops basilaris* (Say) .. .. .. red-shouldered powder-post beetle
*Xylocoris afer (Reuter) .. .. .. African store bug
**Xylocoris flavipes* (Reuter) .. .. .. cosmopolitan cereal bug
Xylocoris galactinus (Fieber) .. .. hot-bed bug
    Xylodrepa quadripunctata: not *quadripunctata* (Linnaeus, 1758), although misidentified as such in several later published works, see *Dendroxena quadrimaculata*
**Xylopertha crinitarsis* (Imhoff) .. .. (beetle) *larva* = African hairy-legged wood borer

SCIENTIFIC NAMES—Arthropods

Scientific Name   **Common Name**

Xylopertha picea (Olivier) .. .. .. (beetle) *larva* = African spruce wood borer
Xyloperthodes nitidipennis (Murray) .. (beetle) *larva* = West African wood borer
Xylopsocus capucinus (Fabricius) .. .. (beetle) *larva* = Madagascan wood borer
    see also *Xylion adustus*
Xyloterus domesticus(L.) .. .. .. broad-leaved wood ambrosia beetle
    see also *Xyleborus dryographus*
Xyloterus lineatus (Olivier) .. .. .. conifer ambrosia beetle
Xylotrechus colonus (Fabricius) .. .. (beetle) *larva* = rustic borer

    Yezabura: not *Yezabura* Matsumura, 1917, although misidentified as such in several later
        published works, see *Dysaphis*
Yponomeuta cagnagella (Hübner) .. .. spindle ermine moth
    Yponomeuta cognatella Treitschke, see *Yponomeuta cagnagella*
Yponomeuta evonymella (L.) .. .. bird-cherry ermine moth
Yponomeuta malinellus Zeller .. .. apple ermine moth
                                        common small ermine moth (see also
                                              Yponomeuta padella)
Yponomeuta padella (L.) .. .. .. common small ermine moth
    see also *Yponomeuta malinellus*   orchard ermine moth
Yponomeuta rorrella (Hübner) .. .. willow ermine moth
YPONOMEUTIDAE .. .. small ermine moths
    (in particular *Yponomeuta* spp.)

Zabrotes subfasciatus (Boheman) .. .. Mexican bean weevil
Zeiraphera diniana (Guenée) .. .. larch tortrix (moth)
                                              grey larch tortrix
                                              larch bud moth
Zeiraphera ratzeburgiana (Ratzeburg) .. spruce tip tortrix (moth)
Zelotherses, subgenus of *Aphelia*
    Zeuzera aesculi (L.), see *Zeuzera pyrina*
Zeuzera pyrina (L.) .. .. leopard moth
                                              wood leopard moth
†*Zicrona caerulea* (L.) .. .. blue shield bug
                                              blue bug (see also *Dysaphis plantaginea*)
    Zonosema alternata (Fallén), see *Rhagoletis alternata*
Zygina flammigera (Fourcroy) .. .. fruit tree leafhopper
    see also species of *Alnetoidia, Edwardsiana, Ribautiana* and *Typhlocyba*
    Zygina pallidifrons Edwards, see *Hauptidia maroccana*
Zygiobia carpini (Löw) .. .. .. hornbeam leaf gall midge

Scientific—Common Names of Molluscs

Scientific Name **Common Name**

Agriolimax, subgenus of *Deroceras*
   Agriolimax reticulatus (Müller), see *Deroceras reticulatum*
Arion ater (L.) .. .. .. black slug
Arion circumscriptus Johnston ⎫
Arion fasciatus (Nilsson) ⎬ .. .. white-soled slugs
  see also *Arion silvaticus* (Lohmander)
Arion distinctus (Mabille) ⎫
Arion hortensis Férussac ⎬ .. .. .. garden slugs
                                     yellow-soled slug (*Arion hortensis*)
Arion intermedius Normand .. .. hedgehog slug
Arion rufus (L.) .. .. red slug
Arion silvaticus Lohmander .. .. .. silver slug
                                   white-soled slug (see also *Arion circumscriptus* and *Arion fasciatus*)
Arion subfuscus (Draparnaud) .. .. dusky slug

Boettgerilla pallens Simroth .. .. .. worm slug

Carinarion, subgenus of *Arion*
Cepaea hortensis (Müller) .. .. .. smaller banded snail
Cepaea nemoralis (L.) .. .. .. larger banded snail
Cernuella virgata (da Costa) .. .. .. striped snail
Cochlicella acuta (Müller) .. .. .. pointed snail
                                   pointed helicellid
Cornu, subgenus of *Helix*

Deroceras caruanae (Pollonera), see *Deroceras panormitanum*
Deroceras laeve (Müller) .. .. .. marsh slug
Deroceras panormitanum (Lessona & .. chestnut slug
  Pollonera)
Deroceras reticulatum (Müller) .. .. field slug
                                   grey field slug
                                   netted slug

Galba, subgenus of *Lymnaea*

Helicella itala (L.) .. .. heath snail
  Helicella virgata (da Costa), see *Cernuella virgata*

SCIENTIFIC NAMES—Molluscs

Scientific Name	Common Name

Helix aspersa Müller .. .. garden snail
  common snail
  Helix hortensis (Müller), see *Cepaea hortensis*
  Helix nemoralis (L.), see *Cepaea nemoralis*
Helix pomatia L. .. .. Roman snail
  apple snail
  Hygromia striolata (Pfeiffer), see *Trichia striolata*

Kobeltia, subgenus of *Arion*

Limacus, subgenus of *Limax*
Limax flavus L. .. .. cellar slug
  dairy slug
  yellow slug
Limax maximus L. .. .. great slug
Limax valentianus Férussac .. .. .. glasshouse slug
  Limnaea truncatula (Müller), see *Lymnaea truncatula*
Lymnaea glabra (Müller) .. .. .. mud snail
  see also *Lymnaea truncatula*
Lymnaea truncatula (Müller) .. .. dwarf pond snail
  mud snail (see also *Lymnaea glabra*)

Mesarion, subgenus of *Arion*
  Milax budapestensis (Hazay), see *Tandonia budapestensis*
Milax gagates (Draparnaud) .. .. keeled slug
  see also *Tandonia budapestensis* and subterranean slug
    Tandonia sowerbyi
  Milax gracilis (Leydig), see *Tandonia budapestensis*
  Milax sowerbyi (Férussac), see *Tandonia sowerbyi*
MOLLUSCA .. .. slugs, snails, shipworms, etc.

Nototeredo norvagicus (Spengler) .. .. common shipworm

PHOLADIDAE .. .. piddocks

Tandonia budapestensis (Hazay) ⎱ .. .. keeled slugs
Tandonia sowerbyi (Férussac) ⎰ subterranean slugs
  see also *Milax gagates* (Draparnaud)
TEREDINIDAE .. .. shipworms
  (in particular *Teredo* spp.)
  Teredo norvagicus Spengler, see *Nototeredo norvagicus* (Spengler)
†*Testacella* spp. .. .. worm-eating slugs
Trichia striolata (Pfeiffer) .. .. .. strawberry snail

Xylophaga dorsalis Turton .. .. .. false shipworm

Common—Scientific Names of Nematodes and Platyhelminths

Common Name   **Scientific Name**

arrowhead worms   { *Toxascaris* spp.
            Toxocara spp.

barber's pole worm, *see* stomach worm
beef tapeworm   *Taenia saginata* Goeze
beet cyst nematode   *Heterodera schachtii* Schmidt
   beet eelworm, *see* beet cyst nematode
bent-grass leaf-gall nematode   *Subanguina graminophila* (T. Goodey) Brzeski
         syn. *Anguina graminophila* (T. Goodey) Christie
bent-grass seed nematode   *Anguina agrostis* (Steinbuch) Filipjev
   black currant eelworm, *see* chrysanthemum nematode
bladder worms   larvae of *Taenia* spp.
brassica cyst nematode   *Heterodera cruciferae* Franklin
   brassica root eelworm, *see* brassica cyst nematode
broad fish tapeworm (of man)   *Diphyllobothrium latum* (L.)
         syn. *Bothriocephalus latus* (L.)
   broad tapeworm (of man), *see* broad fish tapeworm
brown stomach worms   *Ostertagia* spp.
   bud and leaf nematode, *see* leaf nematode (in part) and chrysanthemum nematode (in part)
   bulb eelworm, *see* stem eelworm
*burrowing nematodes   { *Radopholus citropholus* Huettel, Dickson & Kaplan
            Radopholus similis (Cobb) Thorne

   cabbage root eelworm, *see* brassica cyst nematode
cactus cyst nematode   *Cactodera cacti* (Filipjev & Schuurmans Stekhoven) Krall & Krall
         syn. *Heterodera cacti* Filipjev & Schuurmans Stekhoven
   cactus root eelworm, *see* cactus cyst nematode
carrot cyst nematode   *Heterodera carotae* Jones
   carrot root eelworm, *see* carrot cyst nematode
*cat liver fluke   *Opisthorchis felineus* Rivolta
cereal cyst nematode   *Heterodera avenae* Wollenweber
         syn. *Bidera avenae* (Wollenweber) Krall & Krall
         Heterodera maior O. Schmidt misspelling *H. major*
   cereal root eelworm, *see* cereal cyst nematode
cereal root-knot nematode   *Meloidogyne naasi* Franklin

COMMON NAMES — Nematodes and Platyhelminths

Common Name	Scientific Name
chrysanthemum nematode	*Aphelenchoides ritzemabosi* (Schwartz) Steiner & Buhrer syn. *A. ribes* (Taylor) Goodey
clover cyst nematode	*Heterodera trifolii* Goffart
clover root eelworm, *see* clover cyst nematode	
coenurus cysts	larvae of *Taenia multiceps* Leske
cyst eelworms, *see* cyst nematodes	
cyst nematodes	HETERODERINAE
dagger nematodes	*Xiphinema* spp.
*dog heartworm	*Dirofilaria immitis* (Leidy)
dog hookworm	*Uncinaria stenocephala* (Railliet)
dwarf dog tapeworms larvae = hydatid cysts	*Echinococcus* spp.
dwarf tapeworm	*Hymenolepis nana* (Siebold)
ear cockles eelworm, *see* wheat gall nematode	
eelworms, *see* nematodes	
eyeworms	*Thelazia* spp.
*false root-knot nematodes	*Nacobbus* spp.
fern eelworm, *see* leaf nematode	
fescue leaf-gall nematode	*Anguina graminis* (Hardy) Filipjev
fig cyst nematode	*Heterodera fici* Kir'yanova
fig root eelworm, *see* fig cyst nematode	
fish tapeworm, *see* broad fish tapeworm	
flatworms	PLATYHELMINTHES
flower and leaf-gall nematodes	*Anguina* spp. *Subanguina* spp.
flukes	TREMATODA
galeopsis cyst nematode, *see* hemp-nettle cyst nematode	
gapeworms	*Syngamus* spp.
golden nematode, *see* potato cyst nematode, yellow	
grass cyst nematode	*Punctodera punctata* (Thorne) Mulvey & Stone syn. *Heterodera punctata* Thorne
grass root-gall nematode	*Subanguina radicicola* (Greeff) Paramonov syn. *Anguina radicicola* (Greeff) Teploukhova *Ditylenchus radicicolus* (Greef) Filipjev
heartworm (of dog), *see* dog heartworm	
hemp-nettle cyst nematode	*Heterodera galeopsidis* Goffart
hookworms	*Ancylostoma* spp.
hop cyst nematode	*Heterodera humuli* Filipjev
hop root eelworm, *see* hop cyst nematode	
horse pinworm	*Oxyuris equi* (Schrank)
human large roundworm	*Ascaris lumbricoides* L.
hydatid cysts	larvae of *Echinococcus* spp.

COMMON NAMES—Nematodes and Platyhelminths

Common Name	Scientific Name
intestinal threadworm (of pig)	*Strongyloides ransomi* Schwartz & Alicata
intestinal threadworm (of sheep)	*Strongyloides papillosus* (Wedl)
intestinal worms	*Cooperia* spp. *Nematodirus* spp. *Trichostrongylus* spp.
iris bulb nematode, *see* potato tuber nematode	
Javanese root-knot nematode	*Meloidogyne javanica* (Treub) Chitwood
*kidneyworm (of pig)	*Stephanurus dentatus* Diesing
lance nematodes	*Hoplolaimus* spp.
lancet fluke	*Dicrocoelium lanceolatum* (L.) ? syn. *D. dendriticum* Rudolphi
large roundworm (of horse)	*Parascaris equorum* York & Maplestone syn. *Ascaris megalocephala* (Goeze)
leaf nematode *see also* chrysanthemum nematode	*Aphelenchoides fragariae* (Ritzema Bos) Christie syn. *A. olesistus* (Ritzema Bos) Steiner
lemon-shaped cyst nematodes	*Heterodera* spp.
liver fluke	*Fasciola hepatica* L.
lungworms	*Dictyocaulus* spp.
meadow nematodes, *see* root-lesion nematodes	
milfoil cyst nematode, *see* yarrow cyst nematode	
mushroom spawn nematode	*Aphelenchoides composticola* Franklin *Ditylenchus myceliophagus* J. B. Goodey
narcissus bulb and leaf nematode	*Aphelenchoides subtenuis* (Cobb) Steiner & Buhrer
needle nematodes	*Longidorus* spp.
nematodes	NEMATODA, NEMATODEA
nettle cyst nematode	*Heterodera urticae* Cooper
nodular worm	*Oesophagostomum dentatum* (Rudolphi)
northern root-knot nematode	*Meloidogyne hapla* Chitwood
pale potato cyst nematode, *see* potato cyst nematode, white	
pea-nut root-knot nematode	*Meloidogyne arenaria* (Neal) Chitwood
pea cyst nematode	*Heterodera goettingiana* Liebscher
pea root eelworm, *see* pea cyst nematode	
pig lungworms	*Metastrongylus* spp.
pig roundworm	*Ascaris suum* Goeze
pig stomach worm	*Hyostrongylus rubidus* (Hassall & Stiles)
pine wood nematode	*Bursaphelenchus xylophilus* (Steiner & Buhrer) Nickle
pin nematodes	*Paratylenchus* spp.
pinworm (of man)	*Enterobius vermicularis* (L.)

COMMON NAMES—Nematodes and Platyhelminths

Common Name	Scientific Name
pork tapeworm	*Taenia solium* L. syn. *Cysticercus cellulosae* (Gmelin)
potato cyst nematode, white	*Globodera pallida* (Stone) Behrens syn. *Heterodera pallida* Stone misident. *H. rostochiensis* Wollenweber
potato cyst nematode, yellow	*Globodera rostochiensis* (Wollenweber) Behrens syn. *Heterodera rostochiensis* Wollenweber

potato root eelworm, *see* potato cyst nematode

potato tuber nematode	*Ditylenchus destructor* Thorne
poultry nematodes	*Capillaria* spp.
poultry roundworms	{ *Ascaridia galli* (Schrank) { *Heterakis gallinarum* (Schrank) syn. *H. gallinae* Gmelin
poultry tapeworm	*Davainea proglottina* (Davaine) Blanchard

rain worm, *see* thunderworm

ring nematodes	CRICONEMATINAE

root-knot eelworms, *see* root-knot nematodes

root-knot nematodes	*Meloidogyne* spp.
root-lesion nematodes	*Pratylenchus* spp.
round-cyst nematodes	*Globodera* spp.

roundworms, *see* arrowhead worms and nematodes

rumen flukes	*Paramphistomum* spp.
scabious bud nematode	*Aphelenchoides blastophthorus* Franklin

seatworm, *see* pinworm

sheath nematodes	*Hemicycliophora* spp.
sheep lungworm	*Muellerius capillaris* (Müller)
sheep tapeworms	*Moniezia* spp.
southern root-knot nematode	*Meloidogyne incognita* (Kofoid & White) Chitwood
spiral nematodes	{ *Helicotylenchus* spp. { *Rotylenchus* spp.

stem and bulb eelworm, *see* stem nematode

stem nematode	*Ditylenchus dipsaci* (Kühn) Filipjev

stomach flukes, *see* rumen flukes

stomach worm	*Haemonchus contortus* (Rudolphi) Cobb
stubby-root nematodes	{ *Paratrichodorus* spp. { *Trichodorus* spp.
stunt nematodes	{ TYLENCHORHYNCHINAE { MERLINIINAE

swine kidneyworm, *see* kidneyworm

tapeworms	CESTODA
thorny-headed worms	{ *Acanthocephala* spp. { *Polymorphus* spp.

threadworms, *see* nematodes

thunderworm	*Mermis nigrescens* Dujardin

tuber-rot eelworm, *see* potato tuber nematode

COMMON NAMES—Nematodes and Platyhelminths

Common Name	Scientific Name
vinegar eel, *see* vinegar nematode	
vinegar nematode	*Turbatrix aceti* (Müller) Peters
‡wheat gall nematode not found since c.1956	*Anguina tritici* (Steinbuch) Chitwood
whipworms	*Trichuris* spp.
white potato cyst nematode, *see* potato cyst nematode, white	
yarrow cyst nematode	*Globodera achilleae* (Golden & Klindić) Behrens syn. *Heterodera achilleae* Golden & Klindić
yarrow leaf-gall nematode	*Anguina millefolii* Löw) Filipjev
yellow potato cyst nematode, *see* potato cyst nematode, yellow	

Common—Scientific Names of Annelids

Common Name	Scientific Name
aster worms, *see* pot worms	
†brandling	*Eisenia fetida* (Savigny) misspelling *E. foetida* (Savigny)
†chestnut worm	*Lumbricus castaneus* (Savigny)
†cockspur	*Dendrodrilus rubidus* (Savigny) **ssp.** *subrubicunda* (Eisen) syn. *Dendrobaena subrubicunda* (Eisen)
dew worm, *see* lob worm	
†earthworms	LUMBRICIDAE
gilt tail, *see* cockspur	
†green worm	*Allolobophora chlorotica* (Savigny)
†grey worm	*Aporrectodea caliginosa* (Savigny) syn. *A. nocturna* (Evans)
†lob worm	*Lumbricus terrestris* L.
†long worm	*Aporrectodea longa* (Ude)
manure worm, *see* brandling	
nocturnal worm, *see* grey worm	
pot worms	ENCHYTRAEIDAE
purple worm, *see* chestnut worm	

COMMON NAMES—Annelids

Common Name	Scientific Name
†red worm	*Lumbricus rubellus* Hoffmeister
†rosy worm	*Eisenia rosea* (Savigny)
Segmented worms	ANNELIDA

squirrel tail worm, *see* lobworm

true worms, *see* segmented worms
twachel, *see* lob worm

white worms, *see* pot worms

yellow tail, *see* cockspur

Common—Scientific Names of Arthropods

Common Name	Scientific Name
acarine disease mite	*Acarapis woodi* (Rennie)
acorn cup gall cynipid (wasp)	*Andricus quercuscalicis* (Burgsdorf)
acorn moth	*Cydia splendana* (Hübner)
	syn. *Laspeyresia splendana* (Hübner)
acorn weevils	{ *Curculio glandium* Marsham
	Curculio venosus (Gravenhorst)
adelges	ADELGIDAE
*African brown longhorn beetle	*Phryneta leprosa* (Fabricius)
African bollworm, *see* Old World bollworm	
*African carnation tortrix (moth)	*Epichoristodes acerbella* (Walker)
	syn. *E. iocoma* (Meyrick)
	E. ionephela (Meyrick)
*African cotton leafworm, Egyptian cotton leafworm, Mediterranean climbing cutworm	(moth) larva of *Spodoptera littoralis* (Boisduval)
	syn. *Prodenia littoralis* (Boisduval)
	misident. *P. litura* (Fabricius)
	Spodoptera litura (Fabricius)
*African gimlet	(beetle) larva of *Apate terebrans* (Pallas)
*African hairy-legged wood borer	(beetle) larva of *Xylopertha crinitarsis* (Imhoff)
*African horned wood borer	(beetle) larva of *Bostrychoplites cornutus* (Olivier)
*African lyctid (beetle)	*Lyctus africanus* Lesne
*African metallic-green longhorn beetles	{ *Cordylomera spinicornis* (Fabricius)
	Cordylomera suturalis Chevrolat
*African spruce wood borer	(beetle) larva of *Xylopertha picea* (Olivier)
*African store bug	*Xylocoris afer* (Reuter)
afternoon twist moth	*Syndemis musculana* (Hübner)
air-sac mite	*Cytodites nudus* (Vizioli)
alder bead-gall mite	*Eriophyes laevis* (Nalepa)
	syn. *Phytoptus laevis* Nalepa
alder erineum mite	*Acalitus brevitarsus* (Fockeu)
alder froghopper	*Aphrophora alni* (Fallén)
	misident. *A. spumaria* (L.)
alder leaf gall midge	*Dasineura alni* (Löw)
alder sawfly	*Croesus varus* (de Villaret)
alder sucker (psyllid)	*Psylla alni* (L.)
alder wood wasp	*Xiphydria camelus* (L.)
allied shade moth, *see* light grey tortrix	
almond moth, *see* tropical warehouse moth	
ambrosia beetles	{ PLATYPODIDAE
larvae = pin-hole borers	(in particular *Platypus* spp.)
shot-hole borers	*Xyleborus* spp.
American blight aphid, *see* woolly aphid	

COMMON NAMES—Arthropods

Common Name   **Scientific Name**

American cockroach .. .. *Periplaneta americana* (L.)
*American conifer longhorn beetle .. .. *Ergates spiculatus* (LeConte)
American grass thrips, *see* striate thrips
American house dust mite .. .. .. *Dermatophagoides farinae* Hughes
American juniper aphid .. .. .. *Cinara fresai* Blanchard
*American oak bark beetles .. .. .. *Pseudopityophthorus* spp.
American seed beetle, *see* dried bean beetle
*American serpentine leaf miner .. .. (fly) larva of *Liriomyza trifolii* (Burgess in Comstock)
*American spider beetle .. .. .. *Mezium americanum* (Laporte de Castelnau)
*American twig-pruner beetles .. .. { *Elaphidion mucronatum* (Say)
                                                  { *Elaphidion nanum* (Fabricius)
American white moth, *see* fall webworm
andrenas (bees) .. .. *Andrena* spp.
angle-shades moth .. .. *Phlogophora meticulosa* (L.)
Angoumois grain moth .. .. .. *Sitotroga cerealella* (Olivier)
anopheles mosquito, common, *see* malarial anopheles mosquito, common
†ant beetle .. .. *Thanasimus formicarius* (L.)
†ant damsel bug .. .. *Aptus mirmicoides* (Costa)
                                        syn. *Himacerus lativentris* (Boheman)
                                        H. mirmicoides (Costa)
†anthocorid bugs .. .. *Anthocoris* spp.
antirrhinum beetles .. .. .. { *Brachypterolus pulicarius* (L.)
                                    { *Brachypterolus vestitus* (Kiesenwetter)
antler moth .. .. *Cerapteryx graminis* (L.)
                                        syn. *Charaeas graminis* (L.)
antler sawflies .. .. { *Cladius difformis* (Jurine in Panzer)
                                    { *Cladius pectinicornis* (Fourcroy)
ants .. .. FORMICIDAE
ants, bees, sawflies, wasps, etc. .. .. HYMENOPTERA
anystis mites, *see* whirligig mites
†apanteles, common (parasitic wasp) .. *Apanteles glomeratus* (L.)
†apanteles (parasitic wasps) .. .. .. *Apanteles* spp.
†aphid parasitic wasps .. .. .. { *Aphelinus* spp.
                                 { *Aphidius* spp.
aphids .. .. APHIDIDAE
aphids, bugs, hoppers, etc. .. .. .. HEMIPTERA
aphids, hoppers, mealybugs, phylloxeras,
   psyllids, scale insects, whiteflies .. HOMOPTERA
apple and pear bryobia (mite) .. .. *Bryobia rubrioculus* (Scheuten)
                                        misident. *B. praetiosa* Koch (in part)
apple and plum casebearer .. .. .. (moth) larva of *Coleophora spinella* (Schrank)
                                        syn. *C. coracipennella* (Hübner)
                                        C. nigricella (Stephens)
apple aphid, permanent, *see* green apple aphid
apple blossom weevil .. .. *Anthonomus pomorum* (L.)
†apple blossom weevil parasite (wasp) .. *Scambus pomorum* (Ratzeburg)
                                        syn. *Ephialtes pomorum* (Ratzeburg)
                                        Pimpla pomorum Ratzeburg
apple-bud moth, tufted .. .. .. *Platynota idaeusalis* (Walker)
apple bud weevil .. .. *Anthonomus piri* Kollar
                                        syn. *A. cinctus* Redtenbacher
                                        misspelling *A. pyri* Kollar
apple budworm, brown .. .. .. (moth) larva of *Spilonota ocellana* (Denis & Schiffermüller)
apple budworm, spotted .. .. .. (moth) larva of *Hedya dimidioalba* (Retzius)
apple capsid (bug) .. .. *Plesiocoris rugicollis* (Fallén)
†apple capsid, dark green (bug) .. .. *Orthotylus marginalis* Reuter
†apple capsid, red (bug) .. .. .. *Psallus ambiguus* (Fallén)

COMMON NAMES—Arthropods

Common Name	Scientific Name
apple clearwing moth	*Synanthedon myopaeformis* (Borkhausen) syn. *Aegeria myopaeformis* (Borkhausen) *Conopia myopaeformis* (Borkhausen)
apple—dock aphid	*Dysaphis radicola* (Mordvilko)
apple ermine moth	*Yponomeuta malinellus* Zeller
apple foliage weevil	*Magdalis armigera* (Fourcroy)
apple foliage weevil, lesser	*Mecinus pyraster* (Herbst)
*apple fruit fly *larva* = apple maggot	*Rhagoletis pomonella* (Walsh)
apple fruit miner, *larva of* apple fruit moth	
apple fruit moth *larva* = apple fruit miner	*Argyresthia conjugella* Zeller
apple fruit rhynchites (weevil)	*Rhynchites aequatus* (L.) syn. *Caenorhinus aequatus* (L.)
apple—grass aphid	*Rhopalosiphum insertum* (Walker) syn. *R. crataegellum* (Theobald)
apple leaf and bud mite, *see* apple rust mite	
apple leaf-curling midge, *see* apple leaf midge	
apple leaf midge	*Dasineura mali* (Kieffer)
apple leaf miner	(moth) larva of *Lyonetia clerkella* (L.)
apple leaf roller, *larva of* apple moth, light brown	
apple leaf skeletonizer	(moth) larva of *Choreutis pariana* (Clerck) syn. *Anthophila pariana* (Clerck) *Eutromula pariana* Clerck) *Simaethis pariana* (Clerck)
*apple maggot	(fly) larva of *Rhagoletis pomonella* (Walsh)
apple mealybug	*Phenacoccus aceris* (Signoret)
apple moth, light brown *larva* = apple leaf roller	*Epiphyas postvittana* (Walker) syn. *Austrotortrix postvittana* (Walker) *Tortrix postvittana* Walker
apple moth, straw coloured *larva* = apple skin spoiler	*Blastobasis decorella* (Wollaston)
apple pith moth, *see* pith moth	
apple pygmy moth *larva* = crab apple leaf miner	*Stigmella malella* (Stainton) syn. *Nepticula malella* (Stainton)
apple rust mite	*Aculus schlechtendali* (Nalepa) syn. *Phyllocoptes schlechtendali* Nalepa
apple sawfly	*Hoplocampa testudinea* (Klug)
apple skin spoiler, *larva of* apple moth, straw coloured	
apple sucker (psyllid)	*Psylla mali* (Schmidberger)
apple twig cutter (weevil)	*Rhynchites caeruleus* (Degeer)
apple-wood longhorn beetle	*Pogonocherus hispidulus* (Piller & Mitterpacher)
apple worm	(moth) larva of *Cydia pomonella* (L.) syn. *Carpocapsa pomonella* (L.) *Ernarmonia pomonella* (L.) *Laspeyresia pomonella* (L.)
*Arabian red longhorn beetle	*Stromatium fulvum* (Villers)
arabis midge	*Dasineura alpestris* (Kieffer)
archer's dart moth	*Agrotis vestigialis* (Hufnagel)
Argentine ant	*Iridomyrmex humilis* (Mayr)
armoured scales	DIASPIDIDAE (also known as scale insects)
artichoke gall wasp, *see* larch cone gall cynipid	
artichoke tuber aphid, *see* Jerusalem artichoke tuber aphid	
ash bark beetle	*Leperisinus varius* (Fabricius) misident. *Hylesinus fraxini* (Panzer) misspelling *Lepresinus fraxini* (Panzer)
ash bark beetle, lesser	*Hylesinus oleiperda* (Fabricius)
ash bud moth	*Prays fraxinella* (Bjerkander) syn. *P. curtisella* (Donovan)

COMMON NAMES—Arthropods

Common Name	Scientific Name
ash leaf gall sucker (psyllid)	*Psyllopsis fraxini* (L.)
ash leaf sucker (psyllid)	*Psyllopsis fraxinicola* (Förster)
ash midrib pouch-gall midge	*Dasineura fraxini* (Bremi)
	syn. *Dasineura fraxini* (Kieffer)
ash scale	*Pseudochermes fraxini* (Kaltenbach)
	syn. *Apterococcus fraxini* (Newstead)
*Asian cotton leafworm, Asian and Pacific cutworm	(moth) larva of *Spodoptera litura* (Fabricius)
asparagus beetle	*Crioceris asparagi* (L.)
asparagus fly	*Platyparea poeciloptera* (Schrank)
asparagus miner	(fly) larva of *Ophiomyia simplex* (Loew)
	syn. *Agromyza simplex* Loew
	Melanagromyza simplex (Loew)
aspen leaf beetle	*Phytodecta decemnotata* (Marsham)
aspen leaf gall midge	*Harmandia tremulae* (Winnertz)
	syn. *Diplosis tremulae* (Winnertz)
*auger beetle, common	*Sinoxylon anale* Lesne
auricula root aphid	*Thecabius auriculae* (Murray)
	syn. *Pemphigus auriculae* (Murray)
Australian carpet beetle	*Anthrenocerus australis* (Hope)
Australian cockroach	*Periplaneta australasiae* (Fabricius)
Australian spider beetle	*Ptinus tectus* Boieldieu
*Austrian longhorn beetle	*Isotomus speciosus* (Schneider)
autumn fly, *see* face fly	
autumn moth	*Epirrita autumnata* (Borkhausen)
azalea leaf miner	(moth) larva of *Caloptilia azaleella* (Brants)
	syn. *Gracillaria azaleella* Brants
azalea whitefly	*Pealius azaleae* (Baker & Moles)
	syn. *Aleyrodes azaleae* Baker & Moles
bacon beetle	*Dermestes lardarius* L.
badger flea	*Paraceras melis* (Walker)
badger louse	*Trichodectes melis* (Fabricius)
balsam twig aphid	*Mindarus abietinus* Koch
*bamboo longhorn beetle, yellow	*Rhaphuma annularis* (Fabricius)
	syn. *Chlorophorus annularis* (Fabricius)
*banana spider	*Heteropoda venatoria* (L.)
*banded ash borer	(beetle) larva of *Neoclytus caprea* (Say)
banded glasshouse thrips	*Hercinothrips femoralis* (Reuter)
banded mosquito	*Culiseta annulata* (Schrank)
	syn. *Theobaldia annulata* (Schrank)
banded pine weevil, small	*Pissodes castaneus* (Degeer)
	syn. *P. notatus* (Fabricius)
banded pine weevil	*Pissodes pini* (L.)
banded rose sawfly	*Allantus cinctus* (L.)
	syn. *Emphytus cinctus* (L.)
banded-wing flower thrips	*Aeolothrips tenuicornis* Bagnall
banded-wing palm thrips, *see* palm thrips, banded wing	
bank click beetle	*Hypnoidus riparius* (Fabricius)
	syn. *Cryptohypnus riparius* (Fabricius)
larva = bank wireworm	
bank wireworm, *larva of* bank click beetle	
barberry aphid	*Liosomaphis berberidis* (Kaltenbach)
barberry fly, *see* barberry seed fly	
barberry seed fly	*Anomoia purmunda* (Harris)
	syn. *Anomoea permunda* (Harris)
	Phagocarpus permunda (Harris)
	syn. and misspelling *P. permundus* (Harris)

COMMON NAMES—Arthropods

Common Name	Scientific Name
bark beetles	SCOLYTIDAE (in particular *Scolytus* spp.)
bark bugs, *see* flat bugs	
barklice, *see* psocids	
bark psocid	*Cerobasis guestfalica* (Kolbe)
barley flea beetle	*Phyllotreta vittula* Redtenbacher
barn pseudoscorpion	*Chelifer cancroides* (L.)
barred fruit-tree tortrix (moth)	*Pandemis cerasana* (Hübner) syn. *P. ribeana* (Hübner)
barred red moth	*Hylaea fasciaria* (L.) syn. *Ellopia fasciaria* (L.)
bat bug	*Cimex pipistrelli* Jenyns
bat hard tick	*Ixodes vespertilionis* Koch
bat soft tick	*Argas vespertilionis* (Latreille)
bay sucker (psyllid)	*Trioza alacris* Flor
bay-tree scale	*Dynaspidiotus britannicus* (Newstead) syn. *Aspidiotus britannicus* Newstead
bean beetle	*Bruchus rufimanus* Boheman misident. *B. affinis* Froelich
bean beetles, pea beetles	*Bruchus* spp.
bean flower weevil	*Apion vorax* Herbst
bean root aphid	*Smynthurodes betae* Westwood
bean seed beetle, *see* bean beetle	
bean seed flies	{ *Delia florilega* (Zetterstedt) *Delia platura* (Meigen) syn. *D. cilicrura* (Rondani) syn. and misspelling *Hylemyia cilicrura* (Rondani)
bean weevils, clover weevils, pea weevils	*Sitonia* spp.
bed bug	*Cimex lectularius* L.
bedeguar gall wasp *larva* induces gall named 'robin's pin-cushion'	*Diplolepis rosae* (L.) syn. *Rhodites rosae* L.)
beech aphid	*Phyllaphis fagi* (L.)
beech leafhopper	*Fagocyba cruenta* (Herrich-Schäffer) syn. *Typhlocyba cruenta* Herrich-Schäffer *T. douglasi* Edwards
beech leaf miner, *larva of* beech leaf mining weevil	
beech leaf mining weevil *larva* = beech leaf miner	*Rhynchaenus fagi* (L.) syn. *Orchestes fagi* (L.)
beech pouch-gall midges	{ *Hartigiola annulipes* (Hartig) *Mikiola fagi* (Hartig)
beech scale	*Cryptococcus fagisuga* Lindinger syn. *C. fagi* (Baerensprung)
beech seed moth	*Cydia fagiglandana* (Zeller) syn. *Laspeyresia fagiglandana* (Zeller)
beech winter moth, *see* northern winter moth	
bee flies	BOMBYLIIDAE
*bee-hole borer	(moth) larva of *Xyleutes ceramica* (Walker)
bee louse	*Braula coeca* Nitzsch
bee moth, *see* wax moth, bumble-bee	
bees, ants, sawflies, wasps etc.	HYMENOPTERA
beet armyworm	(moth) larva of *Spodoptera exigua* (Hübner) syn. *Laphygma exigua* (Hübner)
beet bug, *see* beet leaf bug	
beet carrion beetle	*Aclypea opaca* (L.) syn. *Blitophaga opaca* (L.)
beet flea beetle, *see* mangold flea beetle	
beet fly, *see* mangold fly	
beet lace bug, *see* beet leaf bug	

COMMON NAMES—Arthropods

Common Name | **Scientific Name**

beet leaf bug .. .. *Piesma quadratum* (Fieber)
beet leaf miner .. .. (fly) larva of *Pegomya hyoscyami* (Panzer)
    syn. *P. betae* (Curtis)
beet leaf weevil .. .. *Tanymecus palliatus* (Fabricius)
beetle mites .. .. CRYPTOSTIGMATA
    syn. ORIBATIDA
        ORIBATEI
beetles (includes weevils) .. .. .. COLEOPTERA
beet webworm .. .. (moth) larva of *Margaritia sticticalis* (L.)
    syn. *Loxostege sticticalis* (L.)
    beetworm (*larva*), *see* silver y moth
    begonia mite, *see* cyclamen mite
begonia thrips .. .. *Scirtothrips longipennis* (Bagnall)
bellflower gall weevil .. .. *Miarus campanulae* (L.)
*belted chion beetle .. .. *Knulliana cinctus* (Drury)
    syn. *Chion cinctus* (Drury)
Benson's larch sawfly .. .. *Anoplonyx destructor* Benson
    berberis aphid, *see* barberry aphid
    berberis seed fly, *see* barberry seed fly
big-beaked plum mite .. .. *Diptacus gigantorhynchus* (Nalepa)
    syn. *Epitrimerus gigantorhynchus* (Nalepa)
    big bud mite, *see* black currant gall mite
bilberry tortrix (moth) .. .. .. *Aphelia viburnana* (Denis & Schiffermüller)
    syn. *Tortrix viburnana* Denis & Schiffermüller
    birch aphid, *see* downy birch aphid
birch bark beetle .. .. *Scolytus ratzeburgi* Janson
birch bug .. .. *Elasmostethus interstinctus* (L.)
birch leaf roller weevil .. .. .. *Deporaus betulae* (L.)
birch sawflies .. .. { *Pristiphora alpestris* (Konow)
    Pristiphora pseudocoactula (Lindqvist)
    Pristiphora quercus (Hartig)
    Pristiphora testacea (Jurine)
bird-cherry aphid .. .. *Rhopalosiphum padi* (L.)
bird-cherry ermine moth .. .. .. *Yponomeuta evonymella* (L.)
    bird-cherry—oat aphid, *see* bird-cherry aphid
bird ked mites .. .. EPIDERMOPTIDAE
bird parasitic flies .. .. *Ornithomya* spp.
biscuit beetle .. .. *Stegobium paniceum* (L.)
    syn. *Sitodrepa panicea* (L.)
    bishop bug, *see* tarnished plant bug
bishop's mitre (bug) .. .. *Aelia acuminata* (L.)
    biting house fly, *see* stable fly
biting lice, chewing lice *and* sucking lice .. PHTHIRAPTERA
biting lice .. .. RHYNCHOPHTHIRINA
    (*see also* chewing lice)
    syn. MALLOPHAGA (in part)
biting midges .. .. CERATOPOGONIDAE
    (in particular *Culicoides* spp.)
black ant, common .. .. *Lasius niger* (L.)
black ants, large .. .. { *Formica fusca* L.
    Formica lemani Bondroit
†black apple capsid (bug) .. .. .. *Atractotomus mali* (Meyer-Duer)
black arches moth .. .. *Lymantria monacha* (L.)
    syn. *Liparis monacha* (L.)
    black-arched tussock, *see* black arches moth
black bean aphid .. .. *Aphis fabae* Scopoli
    blackbeetle, *see* cockroach, common
    blackberry aphid, *see* blackberry—cereal aphid
blackberry aphid, permanent .. .. *Aphis ruborum* (Börner)

COMMON NAMES—Arthropods

Common Name	Scientific Name
blackberry aphid, scarce	*Macrosiphum funestum* (Macchiati)
	syn. *M. rubifolium* Theobald
blackberry—cereal aphid	*Sitobion fragariae* (Walker)
	syn. *Macrosiphum fragariae* (Walker)
	M. rubiellum Theobald
blackberry flower midge	*Contarinia rubicola* Kieffer
	syn. *C. rubicola* Rübsaamen
blackberry leaf midge	*Dasineura plicatrix* (Loew)
blackberry mite	*Acalitus essigi* (Hassan)
	syn. *Aceria essigi* (Hassan)
	Phytoptus essigi (Hassan)
blackberry stem gall midge	*Lasioptera rubi* Heeger
*black borer	(beetle) larva of *Apate monachus* Fabricius
blackbottle (fly)	*Phormia terraenovae* Robineau-Desvoidy
black carpet beetle	*Attagenus unicolor* (Brahm)
	syn. *A. megatoma* (Fabricius)
	A. piceus (Olivier)
*black cherry fruit fly	*Rhagoletis fausta* (Osten Sacken)
black currant aphid	*Cryptomyzus galeopsidis* (Kaltenbach)
	syn. *Capitophorus galeopsidis* (Kaltenbach)
black currant gall mite	*Cecidophyopsis ribis* (Westwood)
	syn. *Eriophyes ribis* (Westwood)
	Phytoptus ribis Westwood
black currant leaf-curling midge, *see* black currant leaf midge	
black currant leaf midge	*Dasineura tetensi* (Rübsaamen)
black currant sawfly	*Nematus olfaciens* Benson
black cutworm	(moth) larva of *Agrotis ipsilon* (Hugnagel)
black dolphin, *see* black bean aphid	
black domestic psocids	*Lepinotus inquilinus* Heyden
	Lepinotus patruelis Pearman
	Lepinotus reticulatus Enderlein
black flea beetle, small	*Phyllotreta aerea* Allard
	? misident. *P. puntulata* (Marsham)
black flies	SIMULIIDAE
blackfly, *see* black bean aphid	
black fly, common	*Simulium reptans* (L.)
black fungus beetles, *see* mealworm beetles, lesser	
black grain stem sawfly	*Trachelus tabidus* (Fabricius)
	syn. *Cephus tabidus* (Fabricius)
†black ground beetle, common	*Nebria brevicollis* (Fabricius)
black-headed fireworm (moth *larva*), *see* holly leaf tier	
†black-kneed capsid (bug)	*Blepharidopterus angulatus* (Fallén)
†black ladybird, minute (beetle)	*Stethorus punctillum* Weise
†black ladybirds (beetles)	*Chilocorus* spp.
black larder beetle	*Dermestes haemorrhoidalis* Küster
black legume aphid, *see* cowpea aphid	
black-lustred ground beetles	*Harpalus* spp.
black millepedes	*Cylindroiulus londinensis* (Leach)
	syn. *C. teutonicus* (Pocock)
(*see also* white-legged black millepede)	
black peach aphid	*Brachycaudus persicae* (Passerini)
	syn. *Anuraphis masseei* (Theobald)
	A. persicae-niger (Smith)
	Brachycaudus persicaecola (Boisduval)
	misident. *B. prunicola* (Kaltenbach)

COMMON NAMES—Arthropods

Common Name	Scientific Name
black pine beetles	*Hylastes angustatus* (Herbst) *Hylastes ater* (Fabricius) 　attribution to author incorrect *H. ater* (Paykull) *Hylastes attenuatus* Erichson *Hylastes brunneus* Erichson *Hylastes opacus* Erichson
black spruce beetle	*Hylastes cunicularius* Erichson
black vine weevil, *see* vine weevil	
black-waved weevil	*Polydrusus undatus* (Fabricius) 　misident. *P. tereticollis* (Degeer)
black willow aphid, *see* willow aphid, black	
bladder pod midge, *see* brassica pod midge	
blinding breeze fly	*Chrysops caecutiens* (L.)
blister beetle	*Lytta vesicatoria* (L.)
bloodworms	(fly) red larvae of CHIRONOMIDAE
blossom beetles	*Meligethes* spp.
blossom fly, *see* fever fly	
blow-flies, *see* bluebottles	
bluebottles	*Calliphora vicina* Robineau-Desvoidy 　syn. *C. eryrthrocephala* (Meigen) *Calliphora vomitoria* (L.)
bluebottles, flesh flies, greenbottles etc.	CALLIPHORIDAE
blue bug, *see* blue shield bug 　and rosy apple aphid	
blue cattle louse	*Solenopotes capillatus* Enderlein
blue-gum sucker, *see* eucalyptus sucker	
blue oat mite, *see* red-legged earth mite	
†blue shield bug	*Zicrona caerulea* (L.)
blue willow leaf beetle	*Phyllodecta vulgatissima* (L.)
blunt snout pillbug (woodlouse)	*Armadillidium nasatum* Budde-Lund
Blyborough tick, *see* bat soft tick	
body louse (of man)	*Pediculus humanus* L. 　syn. *P. corporis* Degeer 　　*P. tabescentium* Alt 　　*P. vestimenti* Nitzsch
booklice, *see* psocids	
booklouse, *see* stored product psocid 　and outhouse psocid	
book pseudoscorpion	*Cheiridium museorum* (Leach)
bordered straw moth	*Heliothis peltigera* (Denis & Schiffermüller)
bordered white moth	*Bupalus piniaria* (L.)
larva = pine looper	
box leaf mining midge	*Monarthropalpus buxi* (Laboulbène)
box sucker (psyllid)	*Psylla buxi* (L.)
*boxwood borers	(beetle) larvae of *Heterobostrychus aequalis* (Waterhouse) *Heterobostrychus brunneus* (Murray)
bracken bug 　(*see also* harvest mite)	*Monalocoris filicis* (L.)
braconids (wasps)	BRACONIDAE
bramble aphid	*Amphorophora rubi* (Kaltenbach)
bramble shoot moth	*Epiblema uddmanniana* (L.) 　syn. *Notocelia uddmanniana* (L.) 　　*Spilonota uddmanniana* (L.)
bramble shoot webber, *larva of* bramble shoot moth	

COMMON NAMES—Arthropods

Common Name	Scientific Name
brassica flower midge	*Gephyraulus raphanistri* (Kieffer)
brassica pod midge	*Dasineura brassicae* (Winnertz)
†brassy ground beetles	*Bembidion* spp.
brassy willow leaf beetle	*Phyllodecta vitellinae* (L.)

bread beetle, *see* biscuit beetle
breeze flies, *see* clegs, gad flies, horse flies, stouts
bright-line brown-eye moth, *see* tomato moth

Brighton wainscot moth	*Oria musculosa* (Hübner)
brindled beauty moth	*Lycia hirtaria* (Clerck)
	syn. *Biston hirtaria* (Clerck)
bristletails	THYSANURA
broad-barred button moth	*Acleris laterana* (Fabricius)
	syn. *A. latifasciana* (Haworth)
	misident. *A. comariana* (Lienig & Zeller)

broad body louse, *see* chicken body louse

broad-horned flour beetle	*Gnatocerus cornutus* (Fabricius)
broad-leaved pinhole borer	(beetle) larva of *Xyleborus dispar* (Fabricius)
	syn. *Anisandrus dispar* (Fabricius)
broad-leaved wood ambrosia beetles	*Xyleborus dryographus* (Ratzeburg)
	syn. *Anisandrus dryographus* (Ratzeburg)
	misspelling *Xyleborus dryophagus* (Ratzeburg)
	Xyloterus domesticus (L.)
	syn. *Trypodendron domesticum* (L.)
broad mite	*Polyphagotarsonemus latus* (Banks)
	syn. *Hemitarsonemus latus* (Banks)
	Tarsonemus latus Banks
broad-nosed grain weevil	*Caulophilus oryzae* (Gyllenhal)
	misident. *C. latinasus* (Say)
broad-nosed weevils	*Barypeithes* spp.
broad-nosed weevil, hairy	*Barypeithes pellucidus* (Boheman)
broad-nosed weevil, smooth	*Barypeithes araneiformis* (Shrank)
bronzed blossom beetles	*Meligethes aeneus* (Fabricius)
	Meligethes viridescens (Fabricius)
broom and clover weevil	*Sitona macularius* (Marsham)
	syn. *S. crinitus* (Herbst)
broom bark beetle, large	*Hylastinus obscurus* (Marsham)
broom bark beetle, small	*Phloeophthorus rhododactylus* (Marsham)

broom brocade moth, *see* broom moth

broom gall midge	*Asphondylia sarothamni* Loew
broom moth	*Ceramica pisi* (L.)
brown ant	*Lasius brunneus* (Latreille)

brown apple budworm (moth *larva*), *see* apple budworm, brown

brown-banded cockroach	*Supella longipalpa* (Fabricius)
	syn. *Phyllodromia supellectilium* (Serville)
	Supella supellectilium (Serville)
brown chafer (beetle)	*Serica brunnea* (L.)

brown chicken louse, *see* chicken louse, brown

brown china mark moth	*Elophila nymphaeata* (L.)
	syn. *Nymphula nymphaeata* (L.)
	syn. and misspelling *Hydrocampa nympheata* (L.)
brown cockroach	*Periplaneta brunnea* Burmeister

brown dog tick, *see* kennel tick

brown-dotted clothes moth	*Niditinea fuscella* (L.)
	syn. *N. fuscipunctella* (Haworth)
	Tinea fuscipunctella Haworth
brown flour mite	*Gohieria fusca* (Oudemans)

COMMON NAMES—Arthropods

Common Name	Scientific Name
brown house moth	*Hofmannophila pseudospretella* (Stainton) syn. *Borkhausenia pseudospretella* (Stainton)
†brown lacewings	*Hemerobius* spp. *Wesmaelius* spp. syn. *Kimminsia* spp.
brown leaf weevil	*Phyllobius oblongus* (L.)
brown mite, see apple and pear bryobia	
brown oak tortrix (moth), see fruit tree tortrices (*in part*) and variegated golden tortrix (*in part*)	
*brown playboy butterfly	*Hypokopelates antalus* (Hopffer) syn. *Deudorix antalus* (Hopffer) *Virachola antalus* (Hopffer)
brown punctured borer	(beetle) larva of *Dryocoetinus villosus* (Fabricius) syn. *Dryocoetes villosus* (Fabricius)
brown scale	*Parthenolecanium corni* (Bouché) syn. *Eulecanium corni* (Bouché) *E. crudum* Green *Lecanium corni* (Bouché) *L. crudum* (Green)
brown soft scale (*see also* nut scale)	*Coccus hesperidum* L. syn. *Lecanium hesperidum* (L.)
brown spider beetle	*Ptinus clavipes* Panzer syn. *P. brunneus* Duftschmid *P. hirtellus* Sturm *P. testaceus* Olivier
brown spruce aphid	*Cinara pilicornis* (Hartig)
brown-tail moth	*Euproctis chrysorrhoea* (L.) syn. *E. phaeorrphoeus* (Haworth) *Nygmia phaeorrhoea* (Donovan) *Nygmia phaeorrhoeus* (Haworth) *Porthesia chrysorrhoea* (L.)
brown wheat mite, see stone mite	
brown willow leaf beetle	*Galerucella lineola* (Fabricius)
buckthorn—potato aphid	*Aphis nasturtii* Kaltenbach misident. *A. rhamni* Boyer de Fonscolombe
bud and rust mites, see gall mites	
bud moth	*Spilonota ocellana* (Denis & Schiffermüller)
larva = brown apple budworm	
buff ermine moth	*Spilosoma lutea* (Hufnagel)
buff-tip moth	*Phalera bucephala* (L.)
bugs	HETEROPTERA
bugs, aphids, hoppers, etc.	HEMIPTERA
bulb and potato aphid	*Rhopalosiphoninus latysiphon* (Davidson)
bulb fly, see narcissus fly, large	
bulb flies, lesser, see narcissus flies, small	
bulb mites	*Rhizoglyphus callae* Oudemans ? misident. *R. echinopus* (Fumouze & Robin (in part)) *Rhizoglyphus robini* Claparède ? misident. *R. echinopus* (Fumouze & Robin (in part))
bulb scale mite	*Steneotarsonemus laticeps* (Halbert) syn. *Tarsonemus approximatus* Banks *T. laticeps* Halbert
bumble-bee wax moth, see wax moth, bumble-bee	
†bumble-bees	*Bombus* spp.

COMMON NAMES—Arthropods

Common Name	Scientific Name
†burnished ground beetle	*Notiophilus biguttatus* (Fabricius)
burrowing bees, *see* andrenas	
burying beetles,	*Nicrophorus* spp.
	misspelling *Necrophorus*
bush-crickets	TETTIGONIIDAE
bush-crickets, crickets and grasshoppers	SALTATORIA
	syn. ORTHOPTERA
butterflies and moths	LEPIDOPTERA
cabbage aphid	*Brevicoryne brassicae* (L.)
†cabbage aphid hover fly	*Scaeva pyrastri* (L.)
†cabbage aphid parasite (wasp)	*Diaeretiella rapae* (M'intosh)
	syn. *Aphidius brassicae* (Marshall in André)
	Diaeretus rapae (Curtis)
cabbage leaf miner	(fly) larva of *Phytomyza rufipes* Meigen
cabbage maggot, *larva of* cabbage root fly	
cabbage moth	*Mamestra brassicae* (L.)
	syn. *Barathra brassicae* (L.)
	B. albidilinea (Haworth)
	Mamestra albidilinea (Haworth)
	Melanchra brassicae (L.)
†cabbage moth parasite (fly)	*Phryxe nemea* (Meigen)
cabbage root fly	*Delia radicum* (L.)
larva = cabbage maggot	syn. *Chortophila brassicae* (Bouché)
	Delia brassicae (Bouché)
	D. brassicae (Weidemann)
	Erioischia brassicae (Bouché)
	E. brassicae (Weidemann)
	Phorbia brassicae (Bouché)
	syn. and misspelling *Hylemyia brassicae* (Bouché)
cabbage seed weevil	*Ceutorhynchus assimilis* (Paykull)
cabbage stem flea beetle	*Psylliodes chrysocephala* (L.)
cabbage stem weevil	*Ceutorhynchus quadridens* (Panzer)
	syn. *C. pallidactylus* (Marsham)
cabbage thrips, *see* field thrips	
cabbage web moth, *see* diamond-back moth	
cabbage white butterflies	*Pieris* spp.
cabbage whitefly	*Aleyrodes proletella* (L.)
	syn. *A. brassicae* Walker
cabinet beetle, large	*Trogoderma inclusum* LeConte
	misident. *T. versicolor* (Creutzer)
cacao moth, *see* warehouse moth	
cacao weevil, *see* coffee bean weevil	
cactus mealybug	*Spilococcus cactearum* McKenzie
	misident. *Pseudococcus mamillariae* (Bouché)
cactus root mealybug, *see* cactus mealybug	
cadelle beetle	*Tenebroides mauritanicus* (L.)
calendula fly	*Napomyza lateralis* (Fallén)
*California red scale	*Aonidiella aurantii* (Maskell)
canary lung mite	*Sternostoma tracheacolum* Lawrence
Canterbury tick, *see* pigeon soft tick	
capricorn beetle, *see* oak longhorn beetle, large	
capsid bugs	MIRIDAE
	syn. CAPSIDAE

COMMON NAMES—Arthropods

Common Name	Scientific Name
carmine spider mite	*Tetranychus cinnabarinus* (Boisduval)
carnation fly	*Delia cardui* (Meigen)
	misident. *D. brunnescens* (Zetterstedt)
	misident. and misspelling *Hylemyia brunnescens* (Zetterstedt)
carnation leaf blotch miner	(fly) larva of *Amauromyza flavifrons* (Meigen)
carnation thrips	*Thrips atratus* Haliday
	syn. *Taeniothrips atratus* (Haliday)
	T. britteni (Bagnall)
carnation tortrix (moth)	*Cacoecimorpha pronubana* (Hübner)
	syn. *Cacoecia pronubana* (Hübner)
	Tortrix pronubana Hübner
carob moth	*Ephestia calidella* Guenée
carpet beetles	*Anthrenus* spp.
larvae = woolly bears	
carrion beetles	SILPHIDAE
carrot and parsnip flat-body moth	*Agonopterix nervosa* (Haworth)
	syn. *Depressaria nervosa* Haworth
carrot aphid, permanent	*Aphis lambersi* (Börner)
carrot fly	*Psila rosae* (Fabricius)
†carrot fly parasite (wasp)	*Chorebus gracilis* (Nees)
	syn. *Dacnusa gracilis* (Nees)
carrot miner	(fly) larva of *Napomyza carotae* Spencer
carrot root aphid	*Pemphigus phenax* Börner & Blunck
carrot sucker (psyllid)	*Trioza apicalis* Förster
casebearers	(moth) larvae of COLEOPHORIDAE
case-bearing clothes moth, *see* clothes moth, case-bearing	
case-making clothes moth, *see* clothes moth, case-bearing	
castor bean tick, *see* sheep tick	
cat biting louse	*Felicola subrostratus* (Burmeister)
	incorrect attribution to author: *F. subrostrata* (Nitzsch)
cat ear mite, *see* ear mange mite	
cat flea	*Ctenocephalides felis* (Bouché)
cat follicle mite	*Demodex cati* Mégnin
cat head mange mite	*Notoedres cati* (Hering)
	syn. *N. cuniculi* (Gerlach)
cat mange mite, *see* cat follicle mite	
cattle biting fly	*Haematobosca stimulans* (Meigen)
	syn. *Haematobia stimulans* (Meigen)
	Siphona stimulans (Meigen)
cattle biting louse	*Bovicola bovis* L.
	syn. *Damalinia bovis* (L.)
	D. scalaris (Nitzsch)
	D. tauri (L.)
cattle follicle mite	*Demodex bovis* Stiles
cattle mange mite, *see* cattle follicle mite	
cattle sweat fly	*Morellia simplex* (Loew)
cattleya 'fly' (wasp), *see* orchidfly	
cattleya scale	*Parlatoria proteus* (Curtis)
cecid midges, *see* gall midges	
celery fly	*Euleia heraclei* (L.)
	syn. *Acidia heraclei* (L.)
	Philophylla heraclei (L.)
celery leaf beetle	*Phaedon tumidulus* (Germar)
cellar beetles	*Blaps* spp.
cellar woodlouse	*Androniscus dentiger* Verhoeff
centipedes	CHILOPODA
	syn. MYRIAPODA (in part)

COMMON NAMES — Arthropods

Common Name	Scientific Name
cereal fly, dusky-winged	*Opomyza germinationis* (L.)
cereal fly, yellow	*Opomyza florum* (Fabricius)
cereal leaf aphid	*Rhopalosiphum maidis* (Fitch)
cereal leaf beetle	*Oulema melanopa* (L.)
	syn. *Lema melanopa* (L.)
cereal leaf miners	(fly) larvae of *Agromyza ambigua* Fallén
	Agromyza nigrella Rondani
	Hydrellia spp.
	Phytomyza nigra Meigen
cereal rust mite	*Abacarus hystrix* (Nalepa)
chafers, dung beetles, etc.	SCARABAEIDAE
chalcids (wasps)	CHALCIDOIDEA
cheese maggot, *see* cheese skipper	
cheese mite	*Tyrolichus casei* (Oudemans)
(*see also* fungal mite *and* mould mite)	syn. *Tyrophagus casei* (Oudemans)
cheese skipper	(fly) larva of *Piophila casei* (L.)
chenopod aphid	*Hayhurstia atriplicis* (L.)
	syn. *Semiaphis atriplicis* (L.)
cherry bark tortrix (moth)	*Enarmonia formosana* (Scopoli)
	syn. *Ernarmonia formosana* (Scopoli)
	E. woeberiana (Denis & Schiffermüller)
	Laspeyresia woeberiana (Denis & Schiffermüller)
cherry beetle mite	*Humerobates rostrolamellatus* Grandjean
	syn. *Euzetes lapidarius* Michael
cherry blackfly	*Myzus cerasi* (Fabricius)
cherry fruit moth	*Argyresthia pruniella* (Clerck)
	misident. *A. curvella* (L.)
	A. nitidella (Fabricius)
cherry leafhopper	*Aguriahana stellulata* (Burmeister)
	syn. *Cicadella stellulata* (Burmeister)
	Eupteroidea stellulata (Burmeister)
	Eupteryx stellulata (Burmeister)
cherry stink bug, *see* forest bug	
*chestnut longhorn beetle	*Tetropium castaneum* (L.)
chewing lice, biting lice *and* sucking lice	PHTHIRAPTERA
	AMBLYCERA
chewing lice	syn. MALLOPHAGA (in part)
(*see also* biting lice)	ISCHNOCERA
	syn. MALLOPHAGA (in part)
chicken body lice, lesser	*Menacanthus cornutus* (Schömmer)
	Menacanthus pallidulus (Neumann)
chicken body louse	*Menacanthus stramineus* (Nitzsch)
	misident. *M. meleagridis* (L.)
chicken feather mite	*Megninia cubitalis* (Mégnin)
chicken fluff louse	*Goniocotes gallinae* (Degeer)
	syn. *G. hologaster* (Nitzsch)
chicken head louse	*Cuclotogaster heterographus* (Nitzsch)
	syn. *C. eynsfordi* (Theobald)
	Lipeurus heterographus Nitzsch
chicken louse, brown	*Goniodes dissimilis* Denny
chicken louse, large	*Goniodes gigas* (Taschenberg)
	syn. *G. abdominalis* (Piaget)
	misident. *G. hologaster* (Nitzsch)
chicken mite	*Dermanyssus gallinae* (Degeer)
chicken shaft louse	*Menopon gallinae* (L.)
	syn. *M. pallidum* (Nitzsch)
	M. trigonocephalum (von Olfers)

COMMON NAMES—Arthropods

Common Name	Scientific Name
chicken wing louse	*Lipeurus caponis* (L.) syn. *L. variabilis* Burmeister
Chilean wood boring beetle	*Pentarthrum huttoni* Wollaston
Chinese scale, *see* San José scale	
chorioptic mange mite	*Chorioptes bovis* (Hering) syn. *C. caprae* (Delafond & Bourguignon) *C. communis* (Cave) *C. cuniculi* Raillet *C. equi* (Gerlach) *C. ovis* (Zürn)
chrysanthemum aphid	*Macrosiphoniella sanborni* (Gillette)
chrysanthemum blotch miner	(fly) larva of *Trypeta zoe* (Meigen) syn. *Euribia zoe* (Meigen) *Spilographa zoe* (Meigen) *Urophora zoe* (Meigen)
chrysanthemum flea beetle	*Longitarsus succineus* (Foudras) misident. *L. laevis* (Duftschmid)
chrysanthemum gall midge	*Rhopalomyia chrysanthemi* (Ahlberg) syn. *Diarthronomyia chrysanthemi* Ahlberg
chrysanthemum leafhopper	*Eupteryx melissae* Curtis syn. *Cicadella melissae* (Curtis)
†chrysanthemum leaf miner parasites (wasps)	*Dacnusa* spp. syn. *Rhizarcha* spp. *Diglyphus* spp.
chrysanthemum leaf miners	(fly) larvae of *Phytomyza horticola* Goureau ? misident. *P. atricornis* Meigen (in part) *Phytomyza syngenesiae* (Hardy) ? misident. *P. atricornis* Meigen (in part)
chrysanthemum leaf mite, *see* chrysanthemum leaf rust mite	
chrysanthemum leaf rust mite	*Epitrimerus alinae* Liro
chrysanthemum midge, *see* chrysanthemum gall midge	
chrysanthemum russet mite, *see* chrysanthemum leaf rust mite	
chrysanthemum stem fly	*Paroxyna misella* (Loew)
chrysanthemum stem mite, *see* chrysanthemum leaf rust mite	
chrysanthemum stool miner	(fly) larva of *Psila nigricornis* Meigen
chrysanthemum thrips	*Thrips nigropilosus* Uzel
cigar casebearer (moth *larva*), *see* apple and plum casebearer	
cigarette beetle	*Lasioderma serricorne* (Fabricius)
cinnabar moth	*Tyria jacobaeae* (L.) syn. *Callimorpha jacobaeae* (L.)
citrophilus mealybug	*Pseudococcus calceolariae* (Maskell) syn. *P. fragilis* Brain *P. gahani* Green
citrus mealybug	*Planococcus citri* (Risso) syn. *Pseudococcus citri* (Risso)
*citrus mussel scale	*Lepidosaphes beckii* (Newman)
*citrus scale	*Lepidosaphes gloverii* (Packard)
clay-coloured weevil	*Otiorhynchus singularis* (L.) syn. *O. picipes* (Fabricius)
clegs, common	*Haematopota crassicornis* Wahlberg *Haematopota pluvialis* (L.)
clegs, gad flies, horse flies	TABANIDAE
click beetle, garden *larva* = garden wireworm	*Athous haemorrhoidalis* (Fabricius)

COMMON NAMES—Arthropods

Common Name	Scientific Name
click beetles	ELATERIDAE
larvae = wireworms	(in particular species of *Agriotes* and *Athous*)
click beetles, common	*Agriotes lineatus* (L.)
larvae = wireworms	*Agriotes obscurus* (L.)
	Agriotes sputator (L.)
click beetles, upland	*Ctenicera* spp.
larvae = upland wireworms	syn. *Corymbites* spp.
cloakers	*Pterostichus* spp.
	syn. *Feronia* spp.
clothes louse, *see* body louse (of man)	
clothes moth, case-bearing	*Tinea pellionella* (L.)
clothes moth, common	*Tineola bisselliella* (Hummel)
larva = webbing clothes maggot	
clothes moth, large pale	*Tinea pallescentella* Stainton
	syn. *Acedes pallescentella* (Stainton)
clothes moth, white-tip	*Trichophaga tapetzella* (L.)
clouded border moth	*Lomaspilis marginata* (L.)
clouded drab moth	*Orthosia incerta* (Hufnagel)
	syn. *Taeniocampa incerta* (Hufnagel)
	T. instabilis (Denis & Schiffermüller)
clover bryobia (mite)	*Bryobia praetiosa* Koch
clover flower midge, *see* clover seed midge	
clover leaf midge	*Dasineura trifolii* (Loew)
	Hypera nigrirostris (Fabricius)
	syn. *Phytonomus nigrirostris* (Fabricius)
clover leaf weevils	*Hypera postica* (Gyllenhal)
	syn. *H. variabilis* (Herbst)
	Phytonomus variabilis (Herbst)
clover mite, *see* clover bryobia	
clover seed midge	*Dasineura leguminicola* (Lintner)
clover seed midge, white	*Dasineura gentneri* Pritchard
clover seed weevil, white	*Apion dichroum* Bedel
	misident. *A. flavipes* (Fabricius)
	Apion apricans Herbst
clover seed weevils	*Apion assimile* Kirby
	Apion trifolii (L.)
	syn. *A. aestivum* Germar
	Sitona humeralis Stephens
clover weevils	*Sitona lepidus* Gyllenhal
	syn. *S. flavescens* (Marsham)
	Sitona puncticollis Stephens
clover weevils, bean weevils, pea weevils	*Sitona* spp.
clover weevils, common	*Sitona hispidulus* (Fabricius)
	Sitona sulcifrons (Thunberg)
	Eudasyphora spp.
	syn. *Dasyphora* spp.
cluster flies	*Ophyra* spp.
(in part sweat flies)	*Pollenia* spp.
	Thaumatomyia spp.
coastal red tick	*Haemaphysalis punctata* Canestrini & Fanzago
	Melolontha hippocastani Fabricius
cockchafers (beetles)	*Melolontha melolontha* (L.)
larvae = white grubs	syn. *M. vulgaris* Fabricius
cockroaches	BLATTODEA
cockroach, common	*Blatta orientalis* L.
cockroaches, etc	DICTYOPTERA

COMMON NAMES—Arthropods

Common Name	Scientific Name
cocksfoot aphid	*Hyalopteroides humilis* (Walker) syn. *H. dactylidis* (Hayhurst)
cocksfoot midges	*Contarinia dactylidis* (Loew) *Contarinia geniculati* (Reuter) syn. *Stenodiplosis geniculati* Reuter (also known as a foxtail midge) *Contarinia merceri* Barnes (also known as a foxtail midge) *Dasineura dactylidis* Metcalf *Sitodiplosis dactylidis* Barnes
cocksfoot moth	*Glyphipterix simpliciella* (Stephens) syn. *G. fischeriella* (Zeller) misspelling *Glyphipteryx simpliciella* (Stephens) misident. *Glyphipterix cramerella* (Fabricius)

codling moth
  larva = apple worm
	Cydia pomonella (L.)
	syn. *Carpocapsa pomonella* (L.)
	    Ernarmonia pomonella (L.)
	    Laspeyresia pomonella (L.)

coffee bean weevil	*Araecerus fasciculatus* (Degeer)
*coffee berry borer	(beetle) larva of *Hypothenemus hampei* (Ferrari) syn. *Stephanoderes hampei* (Ferrari)
coffin fly	*Conicera tibialis* Schmitz
*Colorado beetle	*Leptinotarsa decemlineata* (Say)
columbine leaf miners	(fly) larvae of *Phytomyza aquilegiae* Hardy *Phytomyza minuscula* Goureau
columbine sawfly	*Pristiphora alnivora* (Hartig)

  common anopheles mosquito, *see* malarial anopheles mosquito, common
  common apanteles, *see* apanteles, common
  common auger beetle, *see* auger beetle, common
  common black ant, *see* black ant, common
  common black fly, *see* black fly, common
  †common black ground beetle, *see* black ground beetle, common
  common clegs, *see* clegs, common
  common click beetles, *see* click beetles, common
  common clothes moth, *see* clothes moth, common
  common clover weevils, *see* clover weevils, common
  common cockroach, *see* cockroach, common
  common crane flies, *see* crane flies, common
  common cutworm, *see* cutworm, common
  common earwig, *see* earwig, common
  common flower bug, *see* flower bug, common
  common froghopper, *see* froghopper, common
  common furniture beetle, *see* furniture beetle, common
  common garden centipede, *see* garden centipede, common
  common gnat, *see* gnat, common
  common goat biting louse, *see* goat biting louse, common
  common gooseberry sawfly, *see* gooseberry sawfly, common
  common green capsid, *see* green capsid, common
  common green grasshopper, *see* green grasshopper, common
  common green lacewing, *see* green lacewing, common
  common harvestman, *see* harvestman, common
  common leaf cutter bee, *see* leaf cutter bee, common
  common leaf weevils, *see* leaf weevils, common
  common malarial anopheles mosquito, *see* malarial anopheles mosquito, common

COMMON NAMES—Arthropods

Common Name	Scientific Name

common marbled carpet moth, *see* marbled carpet moth, common
common pillbug, *see* pillbug, common
common rustic moth, *see* rustic moth, common
common scorpion-fly, *see* scorpion-fly, common
common silk moth, *see* silk moth, common
common small ermine moth, *see* ermine moth, common small
common snake-fly, *see* snake-fly, common
common sweat flies, *see* sweat flies, common
common swift moth, *see* garden swift moth
common stiletto fly, *see* stiletto fly, common
common timberman, *see* timberman, common
common wasp, *see* wasp, common
common white wave moth, *see* white wave moth, common
common window gnat, *see* window gnat, common
confused flour beetle .. .. *Tribolium confusum* Jacquelin du Val
conifer ambrosia beetle .. .. .. *Xyloterus lineatus* (Olivier)
    syn. *Trypodendron lineatum* (Olivier)
*conifer auger beetle .. .. *Sinoxylon conigerum* Gerstaecker
†conifer ladybirds (beetles) .. .. .. { *Aphidecta obliterata* (L.)
    { *Exochomus quadripustulatus* (L.)
conifer spinning mite .. .. *Oligonychus ununguis* (Jacobi)
    syn. and attribution to author incorrect
    Paratetranychus ununguis Zachvatkin
  conifer woolly aphids, *see* adelges
copra beetle .. .. *Necrobia rufipes* (Degeer)
copra itch mite .. .. *Cosmoglyphus laarmani* Samsinak
  copra mite, *see* mould mite
cork moth .. .. *Nemapogon cloacella* (Haworth)
    syn. *Tinea cloacella* Haworth
  corn earworm, *see* Old World bollworm
  corn leaf aphid, *see* cereal leaf aphid
corn moth .. .. *Nemapogon granella* (L.)
    syn. *Tinea granella* (L.)
corn-stack bug .. .. *Scolopostethus pictus* (Schilling)
  corn thrips, *see* grain thrips
*cosmopolitan cereal bug .. .. .. *Xylocoris flavipes* (Reuter)
cosmopolitan food mite .. .. .. *Lepidoglyphus destructor* (Schrank)
    syn. *Glycyphagus destructor* (Schrank)
cosmopolitan grain psocid .. .. .. *Lachesilla pedicularia* (L.)
  cosmopolitan nest mite, *see* poultry litter mite
  cotton bollworm, *see* Old World bollworm
*cotton bud thrips .. .. *Frankliniella schultzei* (Trybom)
  cotton whitefly, *see* tobacco whitefly
  cottony cushion scale, *see* fluted scale
  cow-parsnip flat-body, *see* parsnip moth
cowpea aphid .. .. *Aphis craccivora* Koch
cowpea beetles .. .. *Callosobruchus* spp.
  crab apple leaf miner (*larva*), *see* apple pygmy moth
crab louse (of man) .. .. *Pthirus pubis* (L.)
    syn. *P. inguinalis* Leach
    misspelling *P. inquinalis* Leach
crane flies .. .. TIPULIDAE
    (in particular *Tipula* spp.)
crane flies, common .. .. { *Tipula oleracea* L.
  larvae = leatherjackets   { *Tipula paludosa* Meigen (also known as
    marsh crane fly)
  cream-bordered green pea moth, *see* osier green moth
crickets .. .. GRYLLIDAE
crickets, bush-crickets and grasshoppers SALTATORIA
    syn. ORTHOPTERA

COMMON NAMES—Arthropods

Common Name	Scientific Name
croton-bug, *see* German cockroach	
crown flea beetle	*Phyllotreta diademata* Foudras
crustaceans (woodlice, shrimps, etc.)	CRUSTACEA
cuckoo-spit bug	nymph of *Philaenus spumarius* (L.) syn. *P. leucophthalmus* (L.)
cuckoo-spit insects, *see* froghoppers	
cuckoo wasp	*Vespula austriaca* (Panzer)
cuckoo wasps, *see* ruby-tailed wasps	
cucumber mite (*see also* grainstack mite)	*Tyrophagus neiswanderi* Johnston & Bruce
cucumber root maggots, dung maggots, fungus maggots	(fly) larvae of MYCETOPHILIDAE
currant aphid, permanent	*Aphis schneideri* (Borner) misident. *A. varians* Patch
currant blister aphid, *see* black currant aphid	
currant borer, *larva* of currant clearwing moth	
currant bud mite, *see* black currant gall mite	
currant clearwing moth *larva* = currant borer	*Synanthedon tipuliformis* (Clerck) syn. *Aegeria tipuliformis* (Clerck) *Synanthedon salmachus* (L.)
†currant hover fly	*Syrphus ribesii* (L.)
currant—lettuce aphid (*see also* currant—sowthistle aphid *and* gooseberry—sowthistle aphid)	*Nasonovia ribisnigri* (Mosley) syn. *Myzus kaltenbachi* (Schouteden) *Myzus lactucae* (Schrank) *Nasonovia ribicola* (Kaltenbach) misident. *Myzus hieracii* (Kaltenbach)
currant mealybug, *see* citrophilus mealybug	
currant moth, *see* magpie moth	
currant pug moth	*Eupithecia assimilata* Doubleday
currant root aphid	*Eriosoma ulmi* (L.) syn. *Schizoneura ulmi* (L.)
currant shoot borer	(moth) larva of *Lampronia capitella* (Clerck) misspelling *L. capittella* (Clerck)
currant—sowthistle aphid	*Hyperomyzus lactucae* (L.) syn. *Amphorophora cosmopolitana* Mason (in part)
currant stem aphid	*Rhopalosiphoninus ribesinus* (van der Goot)
currant—yellowrattle aphid	*Hyperomyzus rhinanthi* (Schouteden)
cushion scale	*Chloropulvinaria floccifera* (Westwood) syn. *Pulvinaria floccifera* (Westwood)
cutworms and surface caterpillars	(moth) larvae of { *Agrotis* spp. *Euxoa* spp. *Noctua pronuba* (L.) }
cutworm, common	(moth) larva of *Agrotis segetum* (Denis & Schiffermüller)
cyclamen mite	*Phytonemus pallidus* (Banks) **ssp.** *P. pallidus fragariae* (Zimmerman) (strawberry mite) syn. *Steneotarsonemus fragariae* (Zimmerman) *S. pallidus* (Banks) *Tarsonemus fragariae* Zimmerman *T. pallidus* Banks
cyclamen tortrix (moth), *see* straw-coloured tortrix	
cymbidium scale	*Lepidosaphes machili* (Maskell)
cynipids	CYNIPOIDEA
cypress aphid	*Cinara cupressi* (Buckton) syn. *Neochmosis cupressi* (Buckton) *N. tujae* (del Guercio)

COMMON NAMES—Arthropods

Common Name   Scientific Name

   daddy longlegs, *see* crane flies
damp mite .. ..   *Histiostoma feroniarum* (Dufour)
damselflies and dragonflies .. .. ..   ODONATA
damson—hop aphid .. .. ..   *Phorodon humuli* (Schrank)
                                                   syn. *P. pruni* (Scopoli)
   dark flour beetle, *see* flour beetle, dark
   dark fruit tree tortrix (moth), *see* fruit tree tortrix, dark
   dark green apple capsid (bug), *see* apple capsid, dark green
   dark mealworm beetle, *see* mealworm beetle, dark
   dark strawberry tortrix (moth), *see* strawberry tortrix, dark
   dark sword-grass moth, *see* sword-grass moth, dark
dart moths .. ..   { *Agrotis* spp.
                                                   { *Euxoa* spp.
   larvae = cutworms
               surface caterpillars
death's head hawk moth .. .. ..   *Acherontia atropos* (L.)
death watch beetle .. .. ..   *Xestobium rufovillosum* (Degeer)
                                                   syn. *X. tessellatum* (de Villers)
   debris bug, *see* stack bug
December moth .. ..   *Poecilocampa populi* (L.)
                                                   syn. *Eriogaster populi* (L.)
deer flies .. ..   *Chrysops* spp.
deer fly .. ..   *Lipoptena cervi* (L.)
deer nostril fly .. ..   *Cephenemyia auribarbis* (Meigen)
deer warble fly .. ..   *Hypoderma diana* Brauer
delphinium moth .. ..   *Polychrysia moneta* (Fabricius)
   depluming itch mite, *see* depluming mite
depluming mite .. ..   *Neocnemidocoptes gallinae* (Railliet)
                                                  syn. *Knemidokoptes gallinae* (Raillet)
depressed flour beetle .. .. ..   *Palorus subdepressus* (Wollaston)
*desert locust .. ..   *Schistocerca gregaria* (Forskål)
devil's coach-horse beetle .. .. ..   *Staphylinus olens* Müller
diamond-back moth .. .. ..   *Plutella xylostella* (L.)
                                                 syn. *P. maculipennis* (Curtis)
†diamond-back moth parasite (wasp) ..   *Diadegma fenestralis* (Holmgren)
diving water beetles .. .. ..   *Dytiscus* spp.
dock aphid, permanent .. .. ..   *Aphis rumicis* L.
                                                 syn. *A. carbocolor* Gillette
dock miner .. ..   (fly) larva of *Pegomya nigritarsis*
                                                  (Zetterstedt)
dock sawfly .. ..   *Ametastegia glabrata* (Fallén)
                                                 syn. *Taxonus glabratus* (Fallén)
dog biting louse .. ..   *Trichodectes canis* (Degeer)
                                                 syn. *T. latus* Nitzsch
                                                    T. octopunctatus Denny
   dog ear mite, *see* ear mange mite
dog flea .. ..   *Ctenocephalides canis* (Curtis)
dog follicle mite .. ..   *Demodex canis* Leydig
dog fur mite .. ..   *Cheyletiella yasguri* Smiley
   dog itch mite, *see* itch mite
   dog red-mange mite, *see* dog follicle mite
dog sucking louse .. ..   *Linognathus setosus* (von Olfers)
                                                 syn. *L. piliferus* (Burmeister)
dog tick .. ..   *Ixodes canisuga* Johnston
   dogwood aphid, *see* dogwood—grass aphid
dogwood—grass aphid .. .. ..   *Anoecia corni* (Fabricius)
dor beetles .. ..   *Geotrupes* spp.
dot moth .. ..   *Melanchra persicariae* (L.)
                                                 syn. *Mamestra persicariae* (L.)
dotted border moth .. .. ..   *Agriopis marginaria* (Fabricius)

COMMON NAMES—Arthropods

Common Name	Scientific Name
Douglas fir adelges	*Adelges cooleyi* (Gillette) syn. *Gilletteella cooleyi* (Gillette)
Douglas fir seedfly, *see* Douglas fir seed wasp	
Douglas fir seed wasp	*Megastigmus spermotrophus* Wachtl
downy birch aphid	*Euceraphis punctipennis* (Zetterstedt)
dracaena thrips, *see* palm thrips, banded wing	
dragonflies and damselflies	ODONATA
dried bean beetle	*Acanthoscelides obtectus* (Say) syn. *Bruchus obtectus* Say
dried currant moth, *see* tropical warehouse moth	
dried fruit beetles	*Carpophilus* spp.
dried fruit mite	*Carpoglyphus lactis* (L.)
dried fruit moth, *see* carob moth	
drone fly	*Eristalis tenax* (L.)
larva = rat-tailed maggot	
drug store beetle, *see* biscuit beetle	
*dry wood borer	(beetle) larva of *Serropalpus barbatus* (Schaller)
dry wood boring beetles	MELANDRYIDAE
duck wing louse	*Anaticola crassicornis* (Scopoli) syn. *A. anatis* (Fabricius) *A. squalidus* (Nitzsch)
dun flies, *see* clegs, gad flies, horse flies	
dung beetles, chafers, etc.	SCARABAEIDAE
dung beetles	*Aphodius* spp.
dung flies, lesser	SPHAEROCERIDAE syn. BORBORIDAE
dung fly, *see* dung fly, yellow	
dung fly, yellow	*Scathophaga stercoraria* (L.)
dung maggots, cucumber root maggots, fungus maggots	(fly) larvae of MYCETOPHILIDAE
dusky-winged cereal fly, *see* cereal fly, dusky-winged	
dustlice, *see* psocids	
dust mites	PYROGLYPHIDAE
early moth	*Theria primaria* (Haworth) misident. *T. rupicapraria* (Denis & Schiffermüller)
ear canker mite, *see* ear mange mite	
ear mange mite	*Otodectes cynotis* (Hering)
ear mite, *see* ear mange mite	
earwig, common	*Forficula auricularia* L.
earwig, lesser	*Labia minor* (L.)
earwigs	DERMAPTERA
†egg-eating rove beetles	{ *Aleochara bilineata* Gyllenhal *Aleochara bipustulata* (L.)
*Egyptian cotton leafworm, African cotton leafworm, Mediterranean climbing cutworm	(moth) larva of *Spodoptera littoralis* (Boisduval) syn. *Prodenia littoralis* (Boisduval) misident. *P. litura* (Fabricius) *Spodoptera litura* (Fabricius)
eight-toothed spruce bark beetle, smaller, *see* spruce bark beetle, smaller eight-toothed	
elder aphid	*Aphis sambuci* L.
elephant beetle, *see* strawberry blossom weevil	
†eleven-spot ladybird (beetle)	*Coccinella undecimpunctata* L. syn. *C.11-punctata* L.

COMMON NAMES—Arthropods

Common Name	Scientific Name
Elgin shoot moth	*Rhyacionia duplana* (Hübner)
elm balloon-gall aphid	*Eriosoma lanuginosum* (Hartig)
elm bark beetle, large	*Scolytus scolytus* (Fabricius) syn. *S. destructor* Olivier
elm bark beetle, small	*Scolytus multistriatus* (Marsham)
elm—currant aphid, *see* currant root aphid	
†elm gall bug	*Anthocoris gallarumulmi* (Degeer)
elm—grass-root aphid	*Tetraneura ulmi* (L.)
elm leaf aphid, *see* currant root aphid	
elm leaf gall aphid, *see* elm—grass-root aphid	
erica gall midge	*Wachtliella ericina* (Loew)
eriophid mites, *see* gall mites	
ermine moth, apple, *see* apple ermine moth	
ermine moth, common small	*Yponomeuta padella* (L.)
(*see also* apple ermine moth)	
ermine moths, small	YPONOMEUTIDAE (in particular *Yponomeuta* spp.)
*Ethiopian black longhorn beetle	*Mallodon downesi* Hope
*eucalyptus longhorn beetle	*Phoracantha semipunctata* (Fabricius)
larva = phoracantha borer	
eucalyptus sucker	*Ctenarytaina eucalypti* (Maskell)
*European cherry fruit fly	*Rhagoletis cerasi* (L.)
European chicken flea	*Ceratophyllus gallinae* (Schrank)
European chinch bug	*Ischnodemus sabuleti* (Fallén)
European corn borer	(moth) larva of *Ostrinia nubilalis* (Hübner) syn. *Pyrausta nubilalis* (Hübner)
European fowl mite, *see* fowl mite, northern	
European grain moth, *see* corn moth	
European house dust mite	*Dermatophagoides pteronyssinus* (Trouessart)
*European pine longhorn beetle	*Semanotus undatus* (L.)
European pine woolly aphid, *see* Scots pine adelges	
European powder-post beetle	*Lyctus planicollis* LeConte
European rabbit flea	*Spilopsyllus cuniculi* (Dale)
European red mite, *see* fruit tree red spider mite	
European spruce beetle, *see* spruce bark beetle, great	
European spruce sawfly	*Gilpinia hercyniae* (Hartig)
*European vine moth	*Lobesia botrana* (Denis & Schiffermüller)
external bee mite	*Acarapis externus* Morgenthaler
eyed hawk moth	*Smerinthus ocellata* (L.)
eyed ladybird (beetle), *see* pine ladybird	
eye-spotted bud moth, *see* bud moth	
face fly	*Musca autumnalis* Degeer
face louse, *see* sheep sucking louse	
*fall webworm	(moth) larva of *Hyphantria cunea* (Drury)
*false codling moth	*Cryptophlebia leucotreta* (Meyrick)
false furniture beetle	*Ernobius mollis* (L.)
false red spider mites, *see* false spider mites	
false scorpions	PSEUDOSCORPIONES syn. CHELONETHIDA
false spider mites	*Brevipalpus* spp.
false stable fly	*Muscina stabulans* (Fallén)
feather mite, *see* chicken feather mite	
feather mites	ANALGOIDEA
felted beech scale, *see* beech scale	
fern aphid	*Idiopterus nephrelepidis* Davis
fern bug, *see* bracken bug	

COMMON NAMES — Arthropods

Common Name	Scientific Name
fern mite	*Hemitarsonemus tepidariorum* (Warburton)
fern scale	*Pinnaspis aspidistrae* (Signoret)
	syn. *Chionaspis aspidistrae* Signoret
	Hemichionaspis aspidistrae (Signoret)
fern stem borer	(sawfly) larva of *Blasticotoma filiceti* Klug
fern stem sawfly	*Heptamelus ochroleucus* (Stephens)
fern tortrix (moth), *see* straw-coloured tortrix	
fescue aphid	*Metopolophium festucae* (Theobald)
	syn. *Myzus festucae* Theobald
fever fly	*Dilophus febrilis* (L.)
field cricket	*Gryllus campestris* L.
field thrips	*Thrips angusticeps* Uzel
	syn. *T. usemus* Williams
fig mite	*Aceria ficus* (Cotte)
fig moth, *see* raisin moth *and* warehouse moth	
fig mussel scale	*Lepidosaphes conchyformis* (Gmelin in Linnaeus)
	syn. *L. ficus* (Signoret)
fig sucker (psyllid)	*Homotoma ficus* (L.)
figure of eight moth	*Diloba caeruleocephala* (L.)
	syn. *Episema caeruleocephala* (L.)
figwort weevil	*Cionus scrophulariae* (L.)
filbert bud mite	*Phytoptus avellanae* Nalepa
	syn. *Eriophyes avellanae* (Nalepa)
†filbert bud mite parasite (wasp)	*Tetrastichus eriophyes* Taylor
fine gunpowder-mite (springtail), *see* gunpowder-mite, fine	
firebrat (bristletail)	*Thermobia domestica* (Packard)
†five-spot ladybird (beetle)	*Coccinella quinquepunctata* L.
flat-backed millepedes, *see* flat millepedes	
flat bugs	ARADIDAE
flat grain beetle	*Cryptolestes pusillus* (Schoenherr)
	syn. *C. minutus* (Olivier)
	Laemophloeus minutus (Olivier)
flat-headed borers	(beetle) larvae of BUPRESTIDAE
flat millepedes	{ *Brachydesmus superus* Latzel
	{ *Polydesmus angustus* Latzel
flat scarlet mite	*Cenopalpus pulcher* (Canestrini & Fanzago)
flax flea beetle	*Longitarsus parvulus* (Paykull)
	misident. *L. ater* (Fabricius)
flax flea beetle, large	*Aphthona euphorbiae* (Schrank)
*flax thrips	*Thrips lini* Ladureau
(*see also* field thrips)	syn. *T. linarius* Uzel
flax tortrix (moth)	*Cnephasia asseclana* (Denis & Schiffermüller)
larva = poppy leaf roller	syn. *C. interjectana* (Haworth)
	C. virgaureana (Treitschke)
flea beetles	ALTICINAE
	syn. HALTICINAE
	(in particular *Phyllotreta* spp.)
flea beetle, large blue	*Altica lythri* (Aubé)
fleas	SIPHONAPTERA
flesh flies	SARCOPHAGIDAE
	(in particular *Sarcophaga* spp.)
flesh flies, greenbottles, bluebottles	CALLIPHORIDAE
flesh mite, *see* fowl cyst mite	
flies (includes gnats, midges and mosquitoes)	DIPTERA
*Florida red scale	*Chrysomphalus aonidum* (L.)
	syn. *C. ficus* Ashmead
flounced rustic moth	*Luperina testacea* (Denis & Schiffermüller)
flour beetle, dark	*Tribolium destructor* Uyttenboogaart
flour beetles	*Tribolium* spp.

COMMON NAMES—Arthropods

Common Name	Scientific Name
flour mite	*Acarus siro* L.
(*see also* museum mite)	syn. *Aleurobius farinae* (Degeer)
†flower bug, common	*Anthocoris nemorum* (L.)
flower bugs, *see* anthocorid bugs	
flower thrips	*Frankliniella intonsa* (Trybom)
flower weevils	*Apion* spp.
*fluted scale	*Icerya purchasi* Maskell
fly bug, *see* masked hunter bug	
fly-honeysuckle aphid	*Hyadaphis foeniculi* (Passerini)
	syn. *H. coniellum* Theobald
fly-in-the-eye beetle	*Anotylus tetracarinatus* (Block)
follicle mites	DEMODICIDAE
(*see also* itch mites)	(in particular Demodex spp.)
foreign grain beetle	*Ahasverus advena* (Waltl)
forest bug	*Pentatoma rufipes* (L.)
forest fly	*Hippobosca equina* L.
†forest ground beetle	*Carabus nemoralis* Müller
forked red-barred twist (moth), *see* variegated golden tortrix	
†four-spotted burying beetle	*Dendroxena quadrimaculata* (Scopoli)
	misident. *Xylodrepa quadripunctata* (L.)
†fourteen-spot ladybird (beetle)	*Propylea quattuordecimpunctata* (L.)
	syn. *Propylea* 14-*punctata* (L.)
fowl bug, *see* pigeon bug	
fowl cyst mite	*Laminosioptes cysticola* (Vizioli)
fowl mite, northern	*Ornithonyssus sylviarum* (Canestrini & Fanzago)
	syn. *Bdellonyssus sylviarum* (Canestrini & Fanzago)
	Leiognathus sylviarum (Canestrini & Fanzago)
	Liponyssus sylviarum (Canestrini & Fanzago)
*fowl tick	*Argas persicus* (Oken)
fox-coloured sawfly	*Neodiprion sertifer* (Fourcroy)
	syn. *Diprion sertifer* (Fourcroy)
	Lophyrus sertiferus (Fourcroy)
	Neodiprion rufus Latreille
foxglove aphid, *see* glasshouse and potato aphid	
foxtail midges	*Contarinia geniculati* (Reuter)
	syn. *Stenodiplosis geniculati* Reuter (also known as a cocksfoot midge)
	Contarinia merceri Barnes (also known as a cocksfoot midge)
	Dasineura alopecuri (Reuter)
French 'fly', *see* grainstack mite	
frit fly	*Oscinella frit* (L.)
†frit fly parasite (wasp)	*Basalys tritoma* Thomson
	syn. *Loxotropa tritoma* (Thomson)
froghopper, common	*Philaenus spumarius* (L.)
nymph = cuckoo-spit bug	syn. *P. leucophthalmus* (L.)
froghoppers	CERCOPIDAE
frosted orange moth	*Gortyna flavago* (Denis & Schiffermüller)
fruit bark beetle	*Scolytus rugulosus* (Müller)
	incorrect attribution to author *Scolytus rugulosus* Ratzeburg
fruit bark beetle, large	*Scolytus mali* (Bechstein)
	syn. *S. pruni* (Ratzeburg)
fruit flies, large	TEPHRITIDAE
	syn. TRYPETIDAE

COMMON NAMES—Arthropods

Common Name	Scientific Name
fruit flies, small	DROSOPHILIDAE (in particular *Drosophila* spp.)
fruitlet mining tortrix (moth)	*Pammene rhediella* (Clerck)

fruit tree leafhoppers	*Alnetoidia alneti* (Dahlbom)
	    ssp. *A. alneti coryli* (Tollin)
	  syn. *Erythroneura coryli* (Tollin)
	    E. alneti (Dahlbom)
	Edwardsiana avellanae (Edwards)
	  syn. *Typhlocyba avellanae* Edwards
	Edwardsiana crataegi (Douglas)
	  syn. *Typhlocyba crataegi* Douglas
	    T. froggatti Baker
	Edwardsiana hippocastani (Edwards)
	  syn. *Typhlocyba hippocastani* Edwards
	Edwardsiana prunicola (Edwards)
	  syn. *Typhlocyba prunicola* Edwards
	Ribautiana debilis (Douglas)
	  syn. *Typhlocyba debilis* Douglas
	Ribautiana tenerrima (Herrich-Schäffer)
	  syn. *Typhlocyba tenerrima* Herrich-Schäffer
	Typhlocyba quercus (Fabricius)
	Zygina flammigera (Fourcroy)
	  syn. *Erythroneura flammigera* (Geoffroy in Fourcroy)

fruit tree red spider mite	*Panonychus ulmi* (Koch)
	  syn. *Metatetranychus pilosus* (Canestrini & Fanzago)
	    M. ulmi (Koch)
	    Oligonychus ulmi (Koch)
	    Paratetranychus pilosus Canestrini & Fanzago

†fruit tree red spider mite predators (mites)	*Amblyseius finlandicus* (Oudemans) syn. *Typhlodromus finlandicus* (Oudemans) *Typhlodromus pyri* Scheuten
fruit tree thrips, *see* pear thrips	

fruit tree tortrices (moths)
(*see also* fruit tree tortrix, dark, fruit tree tortrix, large *and* marbled orchard tortrix)
	Acleris rhombana (Denis & Schiffermüller)
	  (also known as rhomboid tortrix)
	  syn. *A. contaminana* (Hübner)
	Apotomis spp.
	  syn. *Argyroploce* spp. (in part)
	Archips crataegana (Hübner)
	  (also known as brown oak tortrix)
	  syn. *A. roborana* (Hübner)
	    Cacoecia crataegana (Hübner)
	    C. roborana (Hübner)
	  misident *A. oporana* (L.) (in part)
	    C. oporana (L.) (in part)
	Ditula angustiorana (Haworth)
	  (also known as red-barred tortrix and vine tortrix)
	  syn. *Batodes angustiorana* (Haworth)
	Olethreutes spp.
	  syn. *Argyroploce* spp. (in part)

fruit tree tortrix (moth), dark	*Pandemis heparana* (Denis & Schiffermüller)
fruit tree tortrix (moth), large	*Archips podana* (Scopoli)
fruit-tree wood ambrosia beetle	*Xyleborus saxeseni* (Ratzeburg) syn. *Anisandrus saxeseni* (Ratzeburg)

COMMON NAMES—Arthropods

Common Name	Scientific Name
fungal mites	*Tyrophagus* spp.
fungus beetles	MYCETOPHAGIDAE
fungus flies	MYCETOPHILIDAE
larvae = cucumber root maggots	
dung maggots	
fungus maggots	
fungus gnats, *see* fungus flies	
fungus maggots, cucumber root maggots,	
dung maggots	(fly) larvae of MYCETOPHILIDAE
fur beetle	*Attagenus pellio* (L.)
furniture beetle, *see* furniture beetle, common and furniture beetles	
furniture beetle, common	*Anobium punctatum* (Degeer)
larva = woodworm	syn. *A. domesticum* (Fourcroy)
	misident. *A. striatum* Fabricius
furniture beetles	ANOBIIDAE
larvae = woodworms	
*furniture carpet beetle	*Anthrenus flavipes* Le Conte
furniture mite, *see* house mite	
gad flies, clegs, horse flies	TABANIDAE
gall flies, *see* fruit flies, large	
gall midges	CECIDOMYIIDAE
gall mites	{ ERIOPHYIDAE
	ERIOPHYOIDEA
gall wasps, *see* cynipids	
gall weevils	*Gymnetron* spp.
†garden centipede, common	*Lithobius forficatus* (L.)
garden centipedes	LITHOBIIDAE
(*see also* wire centipedes)	(in particular *Lithobius* spp.)
garden chafer (beetle)	*Phyllopertha horticola* (L.)
garden click beetle, *see* click beetle, garden	
garden dart moth	*Euxoa nigricans* (L.)
	syn. *Agrotis nigricans* (L.)
garden grass veneer moth	*Chrysoteuchia culmella* (L.)
	syn. *Crambus hortuella* (Hübner)
	C. hortuellus (Hübner)
garden pebble moth	*Evergestis forficalis* (L.)
	syn. *Mesographe forficalis* (L.)
	Pionea forficalis (L.)
garden rose tortrix (moth)	*Acleris variegana* (Denis & Schiffermüller)
garden springtail	*Bourletiella hortensis* (Fitch)
	syn. *B. signatus* (Nicolet)
garden swift moth	*Hepialus lupulinus* (L.)
garden symphylan, *see* glasshouse symphylid	
garden wireworm	(beetle) larva of *Athous haemorrhoidalis* (Fabricius)
garden woodlouse	*Porcellio scaber* Latreille
garden woodlouse, grey	*Oniscus asellus* L.
geometer moths	GEOMETRIDAE
larvae = looper caterpillars	
geranium sawfly	*Protemphytus carpini* (Hartig)
	syn. *Ametastegia carpini* (Hartig)
	Emphytus carpini (Hartig)
	misspelling *Protoemphytus carpini* (Hartig)
German cockroach	*Blatella germanica* (L.)

COMMON NAMES—Arthropods

Common Name	Scientific Name
German wasp	*Vespula germanica* (Fabricius)
	syn. *Paravespula germanica* (Fabricius)
	Vespa germanica Fabricius
geum sawfly	{ *Metallus gei* (Brischke)
	Monophadnoides geniculatus (Hartig)
	syn. *Blennocampa geniculata* (Hartig)
ghost moth, *see* ghost swift moth	
ghost swift moth	*Hepialus humuli* (L.)
giant larch aphid	*Cinara kochiana* (Börner)
giant willow aphid, *see* willow aphid, large	
giant wood wasp	*Urocerus gigas* (L.)
	syn. *Sirex gigas* (L.)
gladiolus thrips	*Thrips simplex* (Morison)
	syn. *Taeniothrips gladioli* Moulton & Steinweden
	T. simplex (Morison)
glasshouse and potato aphid	*Aulacorthum solani* (Kaltenbach)
	syn. *Myzus pseudosolani* Theobald
glasshouse camel-cricket	*Tachycines asynamorus* Adelung
glasshouse 'centipede', *see* glasshouse symphylid	
glasshouse leafhopper	*Hauptidia maroccana* (Melichar)
	syn. *Erythroneura pallidifrons* (Edwards)
	Zygina pallidifrons Edwards
glasshouse mealybug	*Pseudococcus affinis* (Maskell)
	syn. *P. obscurus* Essig
	missident. *P. maritimus* (Ehrhorn)
glasshouse millepede	*Oxidus gracilis* (Koch)
	syn. *Paradesmus gracilis* (Koch)
glasshouse millepede, lesser	*Cylindroiulus britannicus* (Verhoeff)
glasshouse orthezia (Homoptera)	*Orthezia insignis* Browne
glasshouse red spider mite, *see* two-spotted spider mite	
glasshouse symphylid	*Scutigerella immaculata* (Newport)
glasshouse thrips	*Heliothrips haemorrhoidalis* (Bouché)
	syn. *H. adonidum* Haliday
glasshouse whitefly	*Trialeurodes vaporariorum* (Westwood)
†glasshouse whitefly parasite (wasp)	*Encarsia formosa* Gahan
glasshouse wing-spot fly	*Scatella tenuicosta* Collin
globular spider beetle	*Trigonogenius globulus* Solier
†glow-worm	(beetle) larva of *Lampyris noctiluca* (L.)
gnat, common	*Culex pipiens* L.
gnats and mosquitoes	CULICIDAE
goat biting louse, common	*Bovicola caprae* (Gurlt)
	syn. *Damalinia caprae* (Gurlt)
	D. climax (Nitzsch in Giebel)
	D. solida (Rudow)
goat follicle mite	*Demodex caprae* Railliet
goat mange mite, *see* goat follicle mite	
goat moth	*Cossus cossus* (L.)
	syn. *C. ligniperda* Fabricius
goat sucking louse	*Linognathus stenopsis* (Burmeister)
	syn. *L. forficulus* (Rudow)
	L. rupicaprae (Rudow)
	L. schistopygus (Nitzsch)
*golden buprestid (beetle)	*Buprestis aurulenta* L.
golden plusia moth, *see* delphinium moth	
golden spider beetle	*Niptus hololeucus* (Faldermann)
*golden twin spot moth	*Chrysodeixis chalcites* (Esper)
	syn. *Plusia chalcites* (Esper)

COMMON NAMES—Arthropods

Common Name	Scientific Name
gold fringe moth	*Hypsopygia costalis* (Fabricius)
larva = hayworm	syn. *Pyralis costalis* (Fabricius)
gold-tail moth, *see* yellow-tail moth	
gooseberry aphid	*Aphis grossulariae* Kaltenbach
gooseberry bryobia (mite)	*Bryobia ribis* Thomas
	misident. *B. praetiosa* Koch (in part)
gooseberry leaf midge	*Dasineura ribicola* (Kieffer)
gooseberry mite, *see* gooseberry bryobia	
gooseberry red spider mite, *see* gooseberry bryobia	
gooseberry sawfly, common	*Nematus ribesii* (Scopoli)
	syn. *Pteronidea ribesii* (Scopoli)
gooseberry sawfly, pale spotted	*Nematus leucotrochus* Hartig
	syn. *Pteronidea leucotrocha* (Hartig)
gooseberry sawfly, small	*Pristiphora pallipes* Lepeletier
gooseberry—sowthistle aphid	*Hyperomyzus pallidus* Hille Ris Lambers
gorse and broom lace bug	*Dictyonota strichnocera* Fieber
gothic moth	*Naenia typica* (L.)
	syn. *Phalaena typica* L.
gout fly	*Chlorops pumilionis* (Bjerkander)
	syn. *C. taeniopus* Meigen
grain aphid	*Sitobion avenae* (Fabricius)
	syn. *Macrosiphum avenae* (Fabricius)
	M. granarium (Kirby)
grain borer, lesser	(beetle) larva of *Rhyzopertha dominica* (Fabricius)
grain itch mite, *see* straw itch mite	
grain mite, *see* flour mite	
grainstack mite	*Tyrophagus longior* (Gervais)
	syn. *Tyroglyphus longior* Gervais
	Tyrophagus tenuiclavus Zachvatkin
grain thrips	*Limothrips cerealium* Haliday
	syn. *L. avenae* Hinds
grain weevil	*Sitophilus granarius* (L.)
granary mite	*Euroglyphus longior* (Trouessart)
	syn. *Dermatophagoides longior* (Trouessart)
grape erineum mite, *see* vine leaf blister mite	
*grape phylloxera	*Daktulosphaira vitifoliae* (Fitch)
	syn. *Dactylosphaera vitifoliae* (Fitch)
	D. vitifolii (Fitch)
	Phylloxera vastatrix (Planchon)
	P. vitifolii (Fitch)
	Viteus vitifoliae (Fitch)
	V. vitifolii (Fitch)
grass and cereal flies	{ *Geomyza tripunctata* Fallén *Opomyza* spp.
grass and cereal mite	*Siteroptes graminum* (Reuter)
	syn. *Pediculopsis graminum* (Reuter)
grass aphid, *see* fescue aphid	
grass flower thrips	*Chirothrips manicatus* Haliday
grass fly	*Meromyza saltatrix* (L.)
grasshoppers and locusts	ACRIDIDAE
grasshoppers, bush-crickets and crickets	SALTATORIA
	syn. ORTHOPTERA
grassland mite	*Tyrophagus similis* Volgin
grass moths	CRAMBINAE
	syn. CRAMBIDAE
(*see also* garden grass veneer moth)	
grass—pear bryobia (mite)	*Bryobia cristata* (Dugès)
	misident. *B. praetiosa* Koch (in part)

COMMON NAMES—Arthropods

Common Name	Scientific Name
grass thrips	*Aptinothrips rufus* (Haliday) syn. *A. lubbocki* (Bagnall) *A. rufus* (Gmelin) *Aptinothrips stylifer* Trybom

greater horntail, *see* giant wood wasp
greater rice weevil, *see* maize weevil
greater wax moth, *see* honeycomb moth
great green bush-cricket, *see* green bush-cricket, great
great spruce bark beetle, *see* spruce bark beetle, great

green apple aphid	*Aphis pomi* Degeer

†green apple capsid, dark (bug), *see* apple capsid, dark green
greenbottle, *see* sheep maggot fly

greenbottles (flies)	*Lucilia* spp.
greenbottles, bluebottles, flesh flies	CALLIPHORIDAE
green budworm	(moth) larva of *Hedya dimidioalba* (Retzius)
green bush-cricket, great	*Tettigonia viridissima* (L.)
green capsid, common (bug)	*Lygocoris pabulinus* (L.) syn. *Lygus pabulinus* (L.)
green cluster fly	*Eudasyphora cyanella* (Meigen) syn. *Dasyphora cyanella* (Meigen) misident. *Pyrellia lasiophthalma* (Macquart)

greenflies, *see* aphids

greenfly predators	(fly) larvae of SYRPHIDAE

green frogflies, *see* green leafhoppers

green grasshopper, common	*Omocestus viridulus* (L.)

green honey moth, *see* wax moth, bumble-bee
greenhouse——, *see* glasshouse——,

†green lacewing, common	*Chrysopa carnea* Stephens
†green lacewings	*Chrysopa* spp.
green leafhoppers	*Empoasca vitis* (Göthe) syn. *Edwardsiana flavescens* (Fabricius) *Empoasca flavescens* (Fabricius) *Empoasca decipiens* Paoli
green leaf weevil	*Phyllobius maculicornis* Germar
green oak tortrix (moth)	*Tortrix viridana* (L.)

larva = oak leaf roller

green pug moth	*Chlorocylstis rectangulata* (L.)
green spruce aphid	*Elatobium abietinum* (Walker) syn. *Aphis abietina* Walker *Neomyzaphis abietina* (Walker)
green-striped fir aphid	*Cinara pectinatae* (Nördlinger)
green-striped spruce bark aphid	*Cinara stroyani* (Pašek)
green tiger beetle	*Cicindela campestris* L.
*green vegetable bug	*Nezara viridula* (L.)

green-veined white butterfly, *see* white butterfly, green-veined
gregarious poplar sawfly, *see* poplar sawfly, gregarious
gregarious spruce sawfly, *see* spruce sawfly, gregarious

grey-coated longhorn beetle	*Anaglyptus mysticus* (L.)
grey dagger moth	*Acronicta psi* (L.) syn. *Apatele psi* (L.)

grey garden woodlouse, *see* garden woodlouse, grey
grey larch tortrix (moth), *see* larch tortrix
grey pine carpet moth, *see* pine carpet moth, grey

grey pine needle aphid	*Schizolachnus pineti* (Fabricius)
grey red-barred tortrix (moth)	*Argyrotaenia ljungiana* (Thunberg) syn. *A. pulchellana* (Haworth)
grey springtail, minute	*Proisotoma minuta* (Tullberg)
grey squirrel flea	*Orchopeas howardi* (Baker) syn. *O. wickhami* (Baker)

COMMON NAMES—Arthropods

Common Name   **Scientific Name**

grey tortrix (moth)	*Cnephasia stephensiana* (Doubleday)
grey trident, *see* grey dagger moth	
grocers' itch mite, *see* cosmopolitan food mite *and* house mite	
ground beetles	CARABINAE
ground beetles and tiger beetles	CARABIDAE
ground-fleas, *see* white blind springtails	
*groundnut borer, *larva of* groundnut bruchid	
*groundnut bruchid (beetle)	*Caryedon serratus* (Olivier)
larva = groundnut borer	syn. *C. fuscus* (Goeze)
	C. gonagra (Fabricius)
ground weevil	*Barynotus obscurus* (Fabricius)
Guernsey carpet beetle	*Anthrenus sarnicus* Mroczkowski
guinea pig fur mite	*Chirodiscoides caviae* Hirst
	syn. *Campylochirus caviae* (Hirst)
guinea pig mite, *see* guinea pig fur mite	
gunpowder-mites (springtails)	{ *Hypogastrura armata* (Nicolet)
	syn. *Achorutes armatus* Nicolet
	Hypogastrura denticulata (Bagnall)
gunpowder-mite, fine (springtail)	*Hypogastrura manubrialis* (Tullberg)
‡gypsy moth	*Lymantria dispar* (L.)
	syn. *Porthetria dispar* (L.)

hairy broad-nosed weevil, *see* broad-nosed weevil, hairy	
hairy cellar beetles, *see* hairy fungus beetles	
hairy fungus beetles	{ *Mycetaea hirta* (Marsham)
	Typhaea stercorea (L.)
hairy ground springtails	*Orchesella* spp.
hairy spider beetle	*Ptinus villiger* Reitter
hairy wood ant, *see* wood ant, hairy	
hard ticks	IXODIDAE
harvesters, *see* harvestmen	
harvestman, common	*Phalangium opilio* L.
harvestmen	OPILIONES
harvestmen, spiders, mites, ticks, etc.	ARACHNIDA
harvest mite	*Neotrombicula autumnalis* (Shaw)
	syn. *Otonyssus autumnalis* (Shaw)
	Trombicula autumnalis (Shaw)
harvest-spiders, *see* harvestmen	
hawk moths	SPHINGIDAE
hawthorn button-top midge	*Dasineura crataegi* (Winnertz)
hawthorn—carrot aphid	*Dysaphis crataegi* (Kaltenbach)
	syn. *Sappaphis crataegi* (Kaltenbach)
hawthorn leaf erineum mite	*Phyllocoptes goniothorax* (Nalepa)
hawthorn moth	*Scythropia crataegella* (L.)
larva = hawthorn webber	
hawthorn—parsley aphid	*Dysaphis apiifolia* (Theobald)
	spp. *petroselini* (Börner)
	syn. *Sappaphis petroselini* (Börner)
hawthorn sawfly	*Trichiosoma tibiale* Stephens
hawthorn spider mite	*Tetranychus viennensis* Zacher
hawthorn stem midge	*Resseliella crataegi* (Barnes)
	syn. *Thomasiniana crataegi* Barnes
hawthorn webber	(moth) larva of *Scythropia crataegella* (L.)
hay-stack bug, *see* corn-stack bug	
hayworm	(moth) larva of *Hypsopygia costalis* (Fabricius)
hazel aphid	*Myzocallis coryli* (Goeze)
hazel aphid, large	*Corylobium avellanae* (Schrank)

COMMON NAMES—Arthropods

Common Name	Scientific Name
hazel nut weevil	*Curculio nucum* (L.)
hazel leaf roller	(weevil) larva of *Apoderus coryli* (L.)
hazel leaf roller weevil	*Byctiscus betulae* (L.)
	syn. *Rhynchites betuleti* (Fabricius)
	misspelling *R. betuli* (Fabricius)
hazel sawfly	*Croesus septentrionalis* (L.)
head louse (of man)	*Pediculus capitis* Degeer
heart and dart moth	*Agrotis exclamationis* (L.)
heather beetle	*Lochmaea suturalis* (Thomson)
heather flea beetle	*Altica ericeti* (Allard)
heather scale	*Eriococcus devoniensis* (Green)
	syn. *Acanthococcus devoniensis* (Green)
heather weevil	*Strophosomus sus* Stephens
	syn. *S. lateralis* (Paykull)
Hebrew character (moth)	*Orthosia gothica* (L.)
hedgehog flea	*Archaeopsylla erinacei* (Bouché)
hedgehog tick	*Ixodes hexagonus* Leach
	misident. *I. autumnalis* Leach
hedgerow and grassland woodlouse	*Philoscia muscorum* (Scopoli)
hemispherical scale	*Saissetia coffeae* (Walker)
	syn. *Lecanium hemisphaericum* Targioni-Tozzetti
	Saissetia hemisphaerica (Targioni-Tozzetti)
henbane flea beetle	*Psylliodes hyoscyami* (L.)
hen flea, *see* European chicken flea	
herbarium beetle	*Dienerella filum* (Aubé)
	syn. *Cartodere filum* (Aubé)
hessian fly	*Mayetiola destructor* (Say)
hide beetles	*Dermestes* spp.
hog louse, *see* pig louse	
hollyhock seed moth	*Pexicopia malvella* (Hübner)
	syn. *Platyedra malvella* (Hübner)
hollyhock weevil	*Apion radiolus* (Marsham)
holly leaf miner	(fly) larva of *Phytomyza ilicis* Curtis
holly leaf tier, *larva of* holly tortrix (moth)	
holly scale, *see* bay-tree scale	
holly tortrix (moth)	*Rhopobota naevana* (Hübner)
larva = holly leaf tier	syn. *Acroclita naevana* (Hübner)
	Rhopobota unipunctana (Haworth)
†honey bee	*Apis mellifera* L.
honeycomb moth	*Galleria mellonella* (L.)
honeylocust gall midge	*Dasineura gleditchiae* (Osten-Sacken)
honey moth, green, *see* wax moth, bumble-bee	
honeysuckle aphid	*Hyadaphis passerinii* (del Guercio)
(*see also* fly-honeysuckle aphid)	syn. *H. lonicerae* Börner
honeysuckle thrips, *see* yellow flower thrips	
honeysuckle whitefly	*Aleyrodes lonicerae* Walker
	syn. *A. fragariae* Walker
hop capsid	*Calocoris fulvomaculatus* (Degeer)
hop-cat	(moth) larva of *Biston betularia* (L.)
hop—damson aphid, *see* damson—hop aphid	
hop-dog	(moth) larva of *Calliteara pudibunda* (L.)
	syn. *Dasychira pudibunda* (L.)
hop gall wasp, *see* larch cone gall cynipid	
hoppers, aphids, bugs, etc.	HEMIPTERA
hoppers, aphids, mealybugs, phylloxeras, psyllids, scale insects, whiteflies	HOMOPTERA
hop flea beetle	*Psylliodes attenuata* (Koch)
hop froghopper, *see* hop leafhopper	

COMMON NAMES — Arthropods

Common Name	Scientific Name
hop leafhopper	*Evacanthus interruptus* (L.)
	syn. *Euacanthus interruptus* (L.)
hop red spider mite, *see* two-spotted spider mite	
hop root weevil	*Plinthus caliginosus* (Fabricius)
	syn. *Epipolaeus caliginosus* (Fabricius)
hop strig maggot, *larva of* hop strig midge	
hop strig midge	*Contarinia humuli* (Theobald)
larva = hop strig maggot	incorrect attribution to author *C. humuli* (Tölg)
hornbeam leaf gall midge	*Zygiobia carpini* (Löw)
hornbeam whitefly	*Asterobemisia carpini* (Koch)
	syn. *Aleurodes ribium* Douglas
	A. rubicola Douglas
	Asterobemisia avellanae (Signoret)
	Asterochiton carpini (Koch)
horn fly	*Haematobia irritans* (L.)
	syn. *Lyperosia irritans* (L.)
	Siphona irritans (L.)
hornet	*Vespa crabro* L.
hornet clearwing, *see* hornet moth	
hornet moth	*Sesia apiformis* (Clerck)
horse biting louse	*Werneckiella equi* (Denny)
	syn. *Damalinia equi* (Denny)
	D. parumpilosa (Piaget)
	D. pilosa (Giebel)
horse bot fly	*Gasterophilus intestinalis* (Degeer)
horse-chestnut scale	*Pulvinaria regalis* Canard
horse flies, clegs, gad flies	TABANIDAE
horse flies, large	{ *Tabanus bovinus* L.
	Tabanus sudeticus Zeller
horse fly, small	*Tabanus bromius* L.
horse follicle mite	*Demodex equi* Railliet
horse foot mange mite, *see* chorioptic mange mite	
horse itch mite, *see* itch mite	
horse mange mite, *see* horse follicle mite	
horse sucking louse	*Haematopinus asini* (L.)
	syn. *H. macrocephalus* (Burmeister)
hot-bed bug	*Xylocoris galactinus* (Fieber)
*hothouse millepede	*Asiomorpha coarctata* (Saussure)
	syn. *Orthomorpha coarctata* (Saussure)
house cricket	*Acheta domesticus* (L.)
	syn. *Gryllulus domesticus* (L.)
house dust mites	*Dermatophagoides* spp.
house fly	*Musca domestica* L.
house fly, lesser	*Fannia canicularis* (L.)
house fly mite	*Macrocheles muscaedomesticae* (Scopoli)
house gnat, *see* gnat, common	
house longhorn beetle	*Hylotrupes bajulus* (L.)
house mite	*Glycyphagus domesticus* (Degeer)
house mouse flea	*Leptopsylla segnis* (Schoenherr)
house spiders	*Tegenaria* spp.
hover flies	SYRPHIDAE
larvae = greenfly predators	(in particular *Syrphus* spp.)
human flea	*Pulex irritans* L.
human follicle mite	*Demodex folliculorum* (Simon)
humble-bees, *see* bumble-bees	

COMMON NAMES—Arthropods

Common Name	Scientific Name
ichneumons	ICHNEUMONIDAE
*Indian ghoon beetle	*Dinoderus brevis* Horn
Indian meal moth	*Plodia interpunctella* (Hübner)
Indian oak borer	(beetle) larva of *Trinophyllum cribratum* Bates
insects	INSECTA syn. HEXAPODA
*insidious flower bug	*Orius insidiosus* (Say)
iris leaf miner	(fly) larva of *Cerodontha ireos* (Goureau) syn. *Dizygomyza iraeos* (Robineau-Desvoidy) *D. ireos* (Goureau)
iris sawfly	*Rhadinoceraea micans* (Klug)
iris thrips	*Frankliniella iridis* (Watson) syn. *Bregmatothrips iridis* Watson *Iridothrips iridis* (Watson)

Isle of Wight disease mite, *see* acarine disease mite

itch mite	*Sarcoptes scabiei* (L.) syn. *S. canis* Gerlach *S. caprae* Fürstenberg *S. equi* Gerlach *S. ovis* Mégnin *S. suis* Gerlach incorrect attribution to author *S. scabiei* (Degeer) misspelling *S. scabei*
itch mites	*Psorergates* spp.
*ivory-marked longhorn beetle	*Eburia quadrigeminata* (Say)
ivy bark beetle	*Kissophagus hederae* (Schmitt)
ivy bryobia (mite)	*Bryobia kissophila* van Eyndhoven misident. *B. praetiosa* Koch (in part)
ivy gall cynipid (wasp)	*Liposthenus latreillei* (Kieffer)

ivy mite, *see* ivy bryobia

ivy whitefly	*Siphoninus immaculatus* (Heeger)
*Japanese beetle	*Popillia japonica* Newman
*Japanese cherry fruit fly	*Euphranta japonica* Ito
*Japanese ghoon beetle	*Dinoderus japonicus* Lesne
Japanese-larch adelges	*Adelges viridana* (Cholodkovsky)
Jerusalem artichoke tuber aphid	*Trama troglodytes* von Heyden
jewel beetles *larvae* = flat-headed borers	BUPRESTIDAE
jumping bugs, shore bugs	SALDIDAE

jumping plant-lice, *see* psyllids
June bug, *see* garden chafer

juniper gall midge	*Oligotrophus juniperinus* (L.)
juniper scales	{ *Carulaspis minima* (Targioni-Tozzetti) ?misident. *C. visci* (Schrank) (in part) *Diaspis caruelii* Targioni-Tozzetti (in part) *Carulaspis juniperi* (Bouché) ?misident. *C. visci* (Schrank) (in part) *Diaspis caruelii* Targioni-Tozzetti (in part) }
juniper webber moth *larva* = juniper webworm	*Dichomeris marginella* (Fabricius)

juniper webworm, *larva of* juniper webber moth

COMMON NAMES—Arthropods

Common Name	Scientific Name
kennel tick	*Rhipicephalus sanguineus* (Latreille)
khapra beetle	*Trogoderma granarium* Everts
kitchen fly	*Drosophila repleta* Wollaston
knotgrass moth	*Acronicta rumicis* (L.)
laboratory stick insect	*Carausius morosus* (Sinéty)
	syn. and incorrect attribution to author *Dixippus morosus* (Brunner)
	incorrect attribution to author: *Carausius morosus* Brunner
laburnum leaf miner	(moth) larva of *Leucoptera laburnella* (Stainton)
lace bugs	TINGIDAE
lacewings	NEUROPTERA
lackey moth	*Malacosoma neustria* (L.)
ladybirds (beetles)	COCCINELLIDAE

ladybird, two-spot (beetle), *see* two-spot ladybird
ladybird, five-spot (beetle), *see* five-spot ladybird
ladybird, seven-spot (beetle), *see* seven-spot ladybird
ladybird, ten-spot (beetle), *see* ten-spot ladybird
ladybird, eleven-spot (beetle), *see* eleven-spot ladybird
ladybird, fourteen-spot (beetle), *see* fourteen-spot ladybird
ladybird, twentytwo-spot (beetle), *see* twentytwo-spot ladybird
ladybird, twentyfour-spot (beetle), *see* twentyfour-spot ladybird

lappet moth	*Gastropacha quercifolia* (L.)
larch adelges	*Adelges laricis* Vallot
	syn. *A. coccineus* (Ratzeburg)
	A. geniculatus (Ratzeburg)
	A. strobilobius (Kaltenbach)
larch aphid	*Cinara cuneomaculata* (del Guercio)
	syn. *C. boerneri* Hille Ris Lambers
larch bark beetle	*Ips cembrae* (Heer)

larch bud moth, *see* larch tortrix

larch casebearer	(moth) larva of *Coleophora laricella* (Hübner)
larch cone gall cynipid (wasp)	*Andricus fecundator* (Hartig)

larch leaf miner, (moth *larva*), *see* larch casebearer

larch longhorn beetle	*Tetropium gabrieli* Weise
	misident. *T. fuscum* (Fabricius)
larch sawflies	⎰ *Pachynematus imperfectus* (Zaddach and Brischke)
	⎱ *Pristiphora wesmaeli* (Tischbein)
larch sawfly, large	*Pristiphora erichsonii* (Hartig)
	syn. *Lygaeonematus erichsonii* (Hartig)
larch sawfly, small	*Pristiphora laricis* (Hartig)
	syn. *Lygaeonematus laricis* (Hartig)

larch shoot borer, *larva of* larch shoot moth

larch shoot moth	*Argyresthia laevigatella* Herrich-Schäffer
larva = larch shoot borer	syn. *A. atmoriella* Bankes
	Blastotere laevigatella (Herrich-Schäffer)
*larch thrips	*Taeniothrips laricivorus* Kratochvíl & Farský
larch tortrix (moth)	*Zeiraphera diniana* (Guenée)
	syn. *Eucosma diniana* (Guenée)
	Steganoptycha diniana (Guenée)
	misident. *Eucosma griseana* (Hübner)

COMMON NAMES—Arthropods

Common Name	Scientific Name
larch webspinner, *see* web-spinning larch sawfly	
larch woolly aphid, *see* larch adelges	
larder beetle, *see* bacon beetle	
large black ants, *see* black ants, large	
large blue flea beetle, *see* flea beetle, large blue	
large broom bark beetle, *see* broom bark beetle, large	
large cabinet beetle, *see* cabinet beetle, large	
large chicken louse, *see* chicken louse, large	
large elm bark beetle, *see* elm bark beetle, large	
large flax flea beetle, *see* flax flea beetle, large	
large fruit bark beetle, *see* fruit bark beetle, large	
large fruit flies, *see* fruit flies, large	
large fruit tree tortrix, *see* fruit tree tortrix, large	
large hazel aphid, *see* hazel aphid, large	
large horse flies, *see* horse flies, large	
large larch sawfly, *see* larch sawfly, large	
large narcissus fly, *see* narcissus fly, large	
large oak longhorn beetle, *see* oak longhorn beetle, large	
large pale clothes moth, *see* clothes moth, large pale	
large pine aphid, *see* pine aphid, large	
large pine sawfly, *see* pine sawfly large	
large poplar longhorn beetle, *see* poplar longhorn beetle, large	
large raspberry aphid, *see* raspberry aphid, large	
large rose sawfly, *see* rose sawfly, large	
larger pale booklouse, *see* pale booklouse, larger	
large striped flea beetle, *see* striped flea beetle, large	
large timberworm, *see* timberworm, large	
large turkey louse, *see* turkey louse, large	
large walnut aphid, *see* walnut aphid, large	
large white butterfly, *see* white butterfly, large	
large willow aphid, *see* willow aphid, large	
large yellow underwing moth, *see* yellow underwing moth, large	
latania scale	*Hemiberlesia lataniae* (Signoret)
late-wheat shoot flies	*Phorbia securis* Tiensuu misident. *P. genitalis* (Schnabl & Dziedzicki) *Phorbia sepia* (Meigen) syn. *Chortophila sepia* (Meigen)
latrine fly	*Fannia scalaris* (Fabricius)
leaf beetles	CHRYSOMELIDAE
leaf-curling plum aphid	*Brachycaudus helichrysi* (Kaltenbach) syn. *Anuraphis helichrysi* (Kaltenbach)
leaf cutter bee, common	*Megachile centuncularis* (L.)
leaf cutter bees	*Megachile* spp.
leafhoppers	CICADELLIDAE syn. JASSIDAE TYPHLOCYBIDAE
leaf insects and stick insects	PHASMIDA
leaf mining weevils	*Rhynchaenus* spp.
leaf rollers	(moth) larvae of TORTRICIDAE
leaf-rolling rose sawfly	*Blennocampa pusilla* (Klug)
leaf weevils, common	*Phyllobius pyri* (L.) *Phyllobius vespertinus* (Fabricius)
leaf weevils	*Phyllobius* spp.
leather beetle	*Dermestes maculatus* Degeer syn. *D. vulpinus* Fabricius
leatherjackets	(fly) larvae of *Tipula oleracea* L. *T. paludosa* Meigen
Leche's twist moth	*Ptycholoma lecheana* (L.)

COMMON NAMES—Arthropods

Common Name	Scientific Name
leek moth	*Acrolepiopsis assectella* (Zeller)
	syn. *Acrolepia assectella* (Zeller)
	misspelling *Acrolepiopis assectella* (Zeller)
leopard moth	*Zeuzera pyrina* (L.)
	syn. *Z. aesculi* (L.)

    lesser apple foliage weevil, *see* apple foliage weevil, lesser
    lesser armyworm, *see* beet armyworm
    lesser ash bark beetle, *see* ash bark beetle, lesser
    lesser bulb flies, *see* narcissus flies, small
    lesser chicken body lice, *see* chicken body lice, lesser
    lesser dung flies, *see* dung flies, lesser
    lesser earwig, *see* earwig, lesser
    lesser glasshouse millepede, *see* glasshouse millepede, lesser
    lesser grain borer, *see* grain borer, lesser
    lesser house fly, *see* house fly, lesser
    lesser mealworm beetles, *see* mealworm beetles, lesser
    lesser mealworms, *see* mealworms, lesser
    lesser pine sawfly, *see* fox-coloured sawfly
    lesser pine shoot beetle, *see* pine shoot beetle, lesser
    lesser rice weevil, *see* rice weevil, lesser
    lesser rose aphid, *see* rose aphid, lesser
    lesser stag beetle, *see* stag beetle, lesser
    lesser strawberry weevils, *see* strawberry weevils, lesser
    lesser wax moth, *see* wax moth, lesser
    lesser willow sawfly, *see* willow sawfly, lesser
    lesser yellow underwing moth, *see* yellow underwing moth, lesser
    lettuce aphid, *see* currant—lettuce aphid

lettuce root aphid	*Pemphigus bursarius* (L.)

    lettuce root fly, *see* chrysanthemum stool miner

lettuce seed fly	*Botanophila gnava* (Meigen)
	syn. *Chortophila gnava* (Meigen)
	Pegohylemyia gnava (Meigen)

    light brown apple moth, *see* apple moth, light brown

light grey tortrix (moth)	*Cnephasia incertana* (Treitschke)
lighthouse-gall midge	*Dasineura glechomae* (Kieffer)
lilac leaf miner	(moth) larva of *Caloptilia syringella* (Fabricius)
	syn. *Gracillaria syringella* (Fabricius)
lily beetle	*Lilioceris lilii* (Scopoli)
	syn. *Crioceris lilii* (Scopoli)

    lily bulb thrips, *see* lily thrips

lily thrips	*Liothrips vaneeckei* Priesner
†lime capsid (bug)	*Orthotylus nassatus* (Fabricius)
lime leaf aphid	*Eucallipterus tiliae* (L.)
lime leaf gall midge	*Didymomyia tiliacea* (Bremi)
	syn. *D. reamuriana* (Loew)
lime leaf-margin gall midges	{ *Dasineura thomasiana* (Kieffer)
	{ *Dasineura tiliamvolvens* (Rübsaamen)
lime leaf-petiole gall midge	*Contarinia tiliarum* (Kieffer)
lime mite	*Eotetranychus tiliarium* (Hermann)
	syn. *Tetranychus tiliarium* (Hermann)

    linden mite, *see* lime mite
    linseed flea beetle, *see* flax flea beetle

little longhorn beetle	*Tetrops praeusta* (L.)
locust bean moth	*Ectomyelois ceratoniae* (Zeller)
	syn. *Myelois ceratoniae* Zeller
	Spectrobates ceratoniae (Zeller)
locusts and grasshoppers	ACRIDIDAE

    loganberry beetle, *see* raspberry beetle

COMMON NAMES—Arthropods

Common Name	Scientific Name
loganberry cane fly	*Pegomya rubivora* (Coquillett) syn. *P. dentiens* (Pandellé)
long-headed flour beetle	*Latheticus oryzae* Waterhouse
longhorn beetles	CERAMBYCIDAE
long-horned grasshoppers, *see* bush-crickets	
long-horned springtail	*Tomocerus longicornis* (Müller)
longicorn beetles, *see* longhorn beetles	
long-legged mushroom mite	*Linopodes motatorius* (L.)
long-legged mushroom mites	*Linopodes* spp.
long-nosed cattle louse	*Linognathus vituli* (L.) syn. *L. tenuirostris* (Burmeister)
*long-tailed blue butterfly	*Lampides boeticus* (L.)
long-tailed mealybug	*Pseudococcus longispinus* (Targioni-Tozzetti) misident. *P. adonidum* (L.)
long-winged thrips, *see* begonia thrips	
looper caterpillars	(moths) larvae of GEOMETRIDAE
lucerne chalcid (wasp)	*Bruchophagus gibbus* (Boheman) syn. *B. funebris* (Howard) *Eurytoma gibba* Boheman
lucerne-flea (springtail)	*Sminthurus viridis* (L.)
lucerne flower midge	*Contarinia medicaginis* Kieffer
lucerne leaf midge	*Jaapiella medicaginis* (Rübsaamen)
lucerne weevils, *see* clover leaf weevils	
lunar hornet moth	*Sesia bembeciformis* (Hübner) syn. *Sphecia bembeciformis* (Hübner)
lupin aphid	*Macrosiphum albifrons* Essig
*Madagascan wood borers	(beetle) larvae of { *Xylion adustus* (Fåhraeus) *Xylopsocus capucinus* (Fabricius) syn. *Apate capucina* (Fabricius)
magpie moth	*Abraxas grossulariata* (L.)
maize weevil	*Sitophilus zeamais* Motschulsky
malarial anopheles mosquito, common	*Anopheles atroparvus* van Thiel misident. *A. maculipennis* Meigen
mangel flea beetle, *see* mangold flea beetle	
mangel fly, *see* mangold fly	
mangold aphid	*Rhopalosiphoninus staphyleae* (Koch) ssp. *tulipaellus* (Theobald) syn. *Hyperomyzus tulipaella* (Theobald)
mangold flea beetle	*Chaetocnema concinna* (Marsham)
mangold fly	*Pegomya hyoscyami* (Panzer) syn. *P. betae* (Curtis)
larva = beet leaf miner	
many plumed moth	*Alucita hexadactyla* (L.) syn. *Orneodes hexadactyla* (L.)
maple bead-gall mite	*Artacris macrorhynchus* (Nalepa)
maple leaf solitary-gall mite	*Aceria macrochelus* (Nalepa) syn. *Eriophyes macrochelus* (Nalepa)
marbled carpet moth, common	*Chloroclysta truncata* (Hufnagel)
marbled minor moth	*Oligia strigilis* (L.) syn. and incorrect attribution to author *Miana strigilis* (Clerck) *Procus strigilis* (Clerck)

COMMON NAMES—Arthropods

Common Name	Scientific Name
marbled orchard tortrix (moth)	*Hedya dimidioalba* (Retzius)
larva = green budworm	syn. *Argyoploce nubiferana* (Haworth)
spotted apple budworm	*A. variegana* (Hübner)
	Hedya nubiferana (Haworth)
marbled single-dot bell moth, *see* holly tortrix	
marble gall wasp	*Andricus kollari* (Hartig)
	syn. *A. circulans* Mayr
	Cynips kollari (Hartig)
March fly	*Bibio hortulanus* (L.)
March moth	*Alsophila aescularia* (Denis & Schiffermüller)
	syn. *Anisopteryx aescularia* (Denis & Schiffermüller)
marsh crane fly, *see* crane flies, common	
†marsh damsel bug	*Dolichonabis limbatus* (Dahlbom)
marsh tick	*Dermacentor reticulatus* (Fabricius)
masked hunter bug	*Reduvius personatus* (L.)
martin bug	*Oeciacus hirundinis* (Lamark)
	syn. *O. hirundinis* (Jenyns)
mason bee	*Osmia rufa* (L.)
mattress dust mite	*Euroglyphus maynei* (Cooreman)
maybugs (beetles), *see* cockchafers	
mayflies	EPHEMEROPTERA
meadow foxtail thrips	*Chirothrips hamatus* Trybom
meal moth	*Pyralis farinalis* (L.)
meal snout moth, *see* meal moth	
mealworm	(beetle) larva of *Tenebrio molitor* L.
mealworm, dark, *larva of* mealworm beetle, dark	
mealworms, lesser, *larvae of* mealworm beetles, lesser	
mealworm beetle, dark	*Tenebrio obscurus* Fabricius
larva = dark mealworm	
mealworm beetles, lesser	{ *Alphitobius diaperinus* (Panzer)
larvae = lesser mealworms	misident. *A. piceus* (Olivier)
	Alphitobius laevigatus (Fabricius)
mealworm beetle, yellow	*Tenebrio molitor* L.
larva = mealworm	
mealybugs	PSEUDOCOCCIDAE
mealybugs, aphids, hoppers, phylloxeras, psyllids, scale insects, whiteflies	HOMOPTERA
mealy cabbage aphid, *see* cabbage aphid	
mealy peach aphid	*Hyalopterus amygdali* (Blanchard)
mealy plum aphid	*Hyalopterus pruni* (Geoffroy)
	misident. *H. arundinis* (Fabricius)
Mediterranean black scale	*Saissetia oleae* (Olivier)
Mediterranean brocade (moth), *see* African cotton leafworm	
Mediterranean carnation leaf roller (moth), *see* carnation tortrix	
*Mediterranean carnation leaf miner	(fly) larva of *Paraphytomyza dianthicola* (Venturi)
*Mediterranean climbing cutworm, African cotton leafworm, Egyptian cotton leafworm	(moth) larva of *Spodoptera littoralis* (Boisduval)
	syn. *Prodenia littoralis* (Boisduval)
	misident. *P. litura* (Fabricius)
	Spodoptera litura (Fabricius)
Mediterranean flour moth	*Ephestia kuehniella* Zeller
	syn. *Anagasta kuehniella* Zeller
*Mediterranean fruit fly	*Ceratitis capitata* (Wiedemann)
*Mediterranean hawthorn weevil	*Otiorhynchus crataegi* Germar
*Mediterranean pulse beetle	*Bruchus ervi* Froelich

COMMON NAMES—Arthropods

Common Name	Scientific Name
melon and cotton aphid	*Aphis gossypii* Glover
merchant grain beetle	*Oryzaephilus mercator* (Fauvel)
Messea's anopheles mosquito	*Anopheles messeae* Falleroni
metallic-blue clerid (beetle)	*Paratillus carus* (Newman)
Mexican bean weevil	*Zabrotes subfasciatus* (Boheman)
*Mexican grain beetle	*Pharaxonotha kirschi* Reitter
milkbottle fly	*Drosophila funebris* (Fabricius)
milkbottle scuttle fly	*Spiniphora bergenstammii* (Mik)
	syn. *Paraspiniphora bergenstammii* (Mik)
millepedes	DIPLOPODA
	syn. MYRIAPODA (in part)
millipedes, *see* millepedes	
mill moth, *see* Mediterranean flour moth	
mint flea beetle	*Longitarsus ferrugineus* (Foudras)
	syn. *L. waterhousei* Kutschera
mint leaf beetle	*Chrysolina menthastri* (Suffrian)
minute black ladybird (beetle), *see* black ladybird, minute	
minute grey springtail, *see* grey springtail, minute	
minute predatory rove beetle, *see* predatory rove beetle, minute	
mistletoe sucker (psyllid)	*Psylla visci* Curtis
mites and ticks	{ ACARI / ACARINA
mites, spiders, ticks, harvestmen, etc.	ARACHNIDA
mole cricket	*Gryllotalpa gryllotalpa* (L.)
mole flea	*Hystrichopsylla talpae* (Curtis)
monkshood aphid	*Delphiniobium junackianum* (Karsch)
	syn. *D. aconiti* (van der Goot)
mosquitoes and gnats	CULICIDAE
moss fly	*Bradysia tritici* (Coquillett)
	misident. *B. pectoralis* (Staeger)
	Sciara pectoralis (Staeger)
moss springtail	*Heteromurus nitidus* (Templeton)
mothering bug, *see* parent bug	
moth flies	PSYCHODIDAE
moths and butterflies	LEPIDOPTERA
mottled arum aphid	*Aulacorthum circumflexum* (Buckton)
	syn. *Myzus circumflexum* (Buckton)
mottled umber moth	*Erannis defoliaria* (Clerck)
	syn. *Hybernia defoliaria* (Clerck)
mottled willow moth, pale	*Caradrina clavipalpis* (Scopoli)
mottled willow moth, small	*Spodoptera exigua* (Hübner)
larva = beet armyworm	syn. *Laphygma exigua* (Hübner)
mould beetles	CRYPTOPHAGIDAE
mould mite	*Tyrophagus putrescentiae* (Schrank)
	syn. *T. castellanii* Hirst
	Tyroglyphus lintneri Osborn
mound ant, *see* yellow meadow ant	
mountain ash aphid	*Dysaphis sorbi* (Kaltenbach)
	syn. *Sappaphis sorbi* (Kaltenbach)
mouse flea, *see* house mouse flea	
mouse fur mite	*Radfordia affinis* (Poppe)
mouse moth	*Amphipyra tragopoginis* (Clerck)
mouse myobiid mite, *see* mouse fur mite	
mulberry moth, *see* fall webworm	
†*mulberry silkworm	(moth) larva of *Bombyx mori* (L.)
murky meal caterpillar	(moth) larva of *Aglossa caprealis* (Hübner)
museum beetle	*Anthrenus museorum* (L.)
museum mite	*Thyreophagus entomophagus* (Laboulbène)
mushroom cecid (fly)	*Heteropeza pygmaea* Winnertz
mushroom flies, *see* mushroom scuttle flies	

COMMON NAMES—Arthropods

Common Name	Scientific Name
mushroom midges	*Henria psalliotae* Wyatt *Lestremia cinerea* Macquart *Mycophila barnesi* Edwards *Mycophila speyeri* (Barnes)
mushroom mite (*see also* fungal mites)	*Tarsonemus myceliophagus* Hussey

mushroom mites, *see* fungal mites
mushroom phorids, *see* mushroom scuttle flies

mushroom sciarid flies	*Bradysia brunnipes* (Meigen) *Lycoriella auripila* (Winnertz) *Lycoriella solani* (Winnertz)
mushroom scuttle flies	*Megaselia bovista* (Gimmerthal) *Megaselia halterata* (Wood) (also known as Worthing phorid) *Megaselia nigra* (Meigen)
mushroom springtails (*see also* gunpowder-mites)	*Xenylla* spp. *Xenylla mucronata* Axelson *Xenylla welchi* Folsom
musk beetle	*Aromia moschata* (L.)
mussel scale	*Lepidosaphes ulmi* (L.) syn. *Mytilaspis pomorum* (Bouché) *Mytilococcus ulmi* (L.)
†mussel scale predatory mite	*Hemisarcoptes malus* (Shimer)
mustard beetles	*Phaedon armoraciae* (L.) *Phaedon cochleariae* (Fabricius) (also known as watercress beetle)

myobiid mites, *see* mouse fur mite *and* rat fur mite

myocoptic mange mite	*Myocoptes musculinus* (Koch)
nail gall mite	*Eriophyes tiliae* (Pagenstecher) syn. *E. gallarumtiliae* (Turpin) *Phytoptus tiliae* Pagenstecher
narcissus flies, small	*Eumerus strigatus* (Fallén) *Eumerus tuberculatus* (Rondani)
narcissus fly, large	*Merodon equestris* (Fabricius)

narrow brown pine aphid, *see* pine aphid, narrow brown
narrow green pine aphid, *see* pine aphid, narrow green

narrow-necked harvest beetle	*Anthicus floralis* (L.)
narrow-winged pug moth	*Eupithecia nanata* (Hübner)
*navel orangeworm	(moth) larva of *Paramyelois transitella* (Walker)

neck louse, *see* chicken head louse
needle-bug, *see* strawberry blossom weevil
needle-nosed hop bug, *see* hop capsid

nettle gall midge	*Dasineura urticae* (Perris)
nettle ground bug	*Heterogaster urticae* (Fabricius)
nettle leaf weevil	*Phyllobius pomaceus* Gyllenhal

New World screw worm (fly larva), *see* screw worm

New Zealand flax mealybug	*Trionymus diminutus* Leonardi
New Zealand wood-boring beetles	*Euophryum confine* (Broun) **Euophryum rufum* (Broun)
ni moth	*Trichoplusia ni* (Hübner) syn. *Plusia ni* (Hübner)
non-biting midges *larvae* (when red) = bloodworms	CHIRONOMIDAE
*North American cherry fruit fly	*Rhagoletis cingulata* (Loew)

northern fowl mite, *see* fowl mite, northern

COMMON NAMES—Arthropods

Common Name	Scientific Name
northern spruce bark beetle, *see* spruce bark beetle, northern	
northern winter moth, *see* winter moth, northern	
Norwegian wasp	*Dolichovespula norwegica* (Fabricius) syn. *Vespula norwegica* (Fabricius)
nose bot fly	*Gasterophilus haemorrhoidalis* (L.)
notoedric itch mite, *see* cat head mange mite	
November moth	*Epirrita dilutata* (Denis & Schiffermüller)
nun moth, *see* black arches moth	
nut bud tortrix (moth)	*Epinotia tenerana* (Denis & Schiffermüller) misident. *Eucosma penkleriana* (Denis & Schiffermüller) *Panoplia penkleriana* (Denis & Schiffermüller)
nut gall mite, *see* filbert bud mite	
nut leaf blister moth	*Phyllonorycter coryli* (Nicelli) syn. *Lithocolletis coryli* Nicelli
nut leaf weevil	*Strophosomus melanogrammus* (Förster) syn. *S. corvli* (Fabricius)
nutmeg moth	*Discestra trifolii* (Hüfnagel)
nut scale	*Eulecanium tiliae* (L.) syn. *E. coryli* (L.) *Lecanium coryli* (L.)
nut weevil, *see* hazel nut weevil	
oat—apple aphid, *see* apple—grass aphid	
oak-apple gall wasp	*Biorhiza pallida* (Olivier) syn. *B. aptera* (Fabricius) *B. quercusterminalis* (Fabricius)
oak bark beetle	*Scolytus intricatus* (Ratzeburg)
oak bark phylloxera	*Moritziella corticalis* (Kaltenbach)
oak bud collared-gall cynipid (wasp)	*Andricus curvator* Hartig
oak bud globular-gall cynipid (wasp)	*Andricus inflator* Hartig
oak bud red-gall cynipid (wasp)	*Cynips divisa* Hartig
oak catkin gall cynipid (wasp)	*Andricus nudus* Adler
†oak flower bug	*Anthocoris confusus* Reuter
oak fold-gall midges	*Macrodiplosis* spp.
oak gall cynipids	*Andricus* spp.
oak gall wasps, *see* oak gall cynipids	
oak leaf aphid	*Tuberculoides annulatus* (Hartig)
oak leaf blister-gall cynipid (wasp)	*Neuroterus numismalis* (Fourcroy)
oak leaf cherry-gall cynipid (wasp)	*Cynips quercusfolii* L.
oak leaf cupped-gall cynipid (wasp)	*Neuroterus tricolor* (Hartig)
oak leaf gall cynipids (wasps)	{ *Andricus albopunctatus* (Schlechtendal) *Andricus quadrilineatus* Hartig *Andricus solitarius* (Fonscolombe)
oak leaf kidney-gall cynipid (wasp)	*Trigonaspis megaptera* (Panzer)
oak leaf mining weevil	*Rhynchaenus quercus* (L.)
oak leaf oyster-gall cynipid (wasp)	*Andricus ostreus* (Hartig) syn. *A. furunculus* (Beyerinck)
oak leaf phylloxera	*Phylloxera glabra* (von Heyden) syn. *P. punctata* Lichtenstein
oak leaf roller	(moth) larva of *Tortrix viridana* (L.)
oak leaf roller weevil	*Attelabus nitens* (Scopoli)
oak leaf sucker (psyllid)	*Trioza remota* Förster
oak leaf smooth-gall cynipid (wasp)	*Neuroterus albipes* (Schenck)
oak leaf spangle-gall cynipid (wasp)	*Neuroterus quercusbaccarum* (L.) syn. *N. lenticularis* (Olivier)

COMMON NAMES—Arthropods

Common Name	Scientific Name
oak leaf striped-gall cynipid (wasp)	*Cynips longiventris* Hartig
oak longhorn beetles	⎰ *Prionus coriarius* (L.) (also known as tanner beetle) ⎱ *Rhagium mordax* (Degeer)
‡oak longhorn beetle, large	*Cerambyx cerdo* L.
oak louse	*Phylloxera quercus* Fonscolombe
oak pin-hole borer	(beetle) larva of *Platypus cylindrus* (Fabricius)
oak pit gallers, *see* oak pit scales	
oak pit scales (Homoptera)	⎰ *Asterodiaspis minus* (Lindinger) ⎨ *Asterodiaspis quercicola* (Bouché) ⎱ *Asterodiaspis variolosa* (Ratzeburg)
oak red barnacle-gall cynipid (wasp)	*Andricus testaceipes* Hartig
oak roller moth, *see* green oak tortrix	
oak root truffle-gall cynipid (wasp)	*Andricus quercusradicis* (Fabricius)
oak scales	⎰ *Eulecanium ciliatum* (Douglas) ⎱ *Kermes quercus* (L.)
oak slugworm	(sawfly) larva of *Caliroa annulipes* (Klug)
oak terminal-shoot gall midge	*Arnoldiola quercus* (Binnie)
oak tortrix (moth), brown, *see* fruit tree tortrices *and* variegated golden tortrix	
oak tortrix (moth), green, *see* green oak tortrix	
oak whitefly	*Pealius quercus* (Signoret) misident. *P. avellanae* (Signoret)
†oakwood ground beetle	*Calosoma inquistor* (L.)
oat—apple aphid, *see* apple—grass aphid	
oat leaf beetle, *see* cereal leaf beetle	
oat spiral mite	*Steneotarsonemus spirifex* (Marchal) syn. *Tarsonemus spirifex* Marchal
oat stem midge	*Mayetiola avenae* (Marchal)
oat thrips	*Stenothrips graminum* Uzel syn. *Baliothrips graminum* (Uzel)
odd beetle	*Thylodrias contractus* Motschulsky
oil beetles	*Meloe* spp.
*Old World bollworm	(moth) larva of *Helicoverpa armigera* (Hübner) syn. *Heliothis armigera* (Hübner) misident. *Chloridea obsoleta* (Fabricius) *Heliothis obsoleta* (Fabricius) *H. zea* (Boddie)
oleander aphid	*Aphis nerii* Boyer de Fonscolombe
oleander scale	*Aspidiotus nerii* Bouché misident. *A. hederae* (Vallot)
omnivorous leaf tier	(moth) larva of *Cnephasia longana* (Haworth)
onion fly	*Delia antiqua* (Meigen) syn. and misspelling *Hylemyia antiqua* (Meigen)
onion thrips	*Thrips tabaci* Lindeman syn. *T. debilis* Bagnall
open ground symphylids	*Symphylella* spp.
orange wheat blossom midge, *see* wheat blossom midge, orange	
orchard ermine moth, *see* apple ermine moth	
orchid aphid	*Cerataphis orchidearum* (Westwood)
orchid aphid, yellow	*Sitobion luteum* (Buckton)
orchidfly (wasp)	*Eurytoma orchidearum* (Westwood)
orchid scale	*Diaspis boisduvalii* Signoret
orchid springtails, *see* hairy ground springtails	
orchid thrips, yellow	*Anaphothrips orchidaceus* Bagnall
oribatid mites, *see* beetle mites	

COMMON NAMES—Arthropods

Common Name	Scientific Name
oriental cockroach, *see* cockroach, common	
*oriental fruit moth	*Cydia molesta* (Busck)
	syn. *Grapholita molesta* (Busck)
oriental pine adelges	*Pineus orientalis* (Dreyfus)
oriental rat flea	*Xenopsylla cheopis* (Rothschild)
osier green moth	*Earias clorana* (L.)
osier leaf-folding midge	*Rhabdophaga marginemtorquens* (Bremi)
(*see also* willow leaf-folding midge)	
osier weevil	*Cryptorhynchus lapathi* (L.)
larva = willow borer	
outhouse psocid	*Liposcelis subfuscus* Broadhead
	?syn. *L. divinatorius* (Müller) (in part)
owl midges, *see* moth flies	
ox bot flies, *see* ox warble flies	
ox psoroptic mange mite, *see* psoroptic mange mite	
ox tail mange mite, *see* chorioptic mange mite	
ox warble flies	{ *Hypoderma bovis* (L.)
	Hypoderma lineatum (Villers)
oyster scale, *see* oystershell scales	
oystershell scales	{ *Quadraspidiotus ostreaeformis* (Curtis) (also known as oyster scale)
	syn. *Aspidiotus ostreaeformis* Curtis
	Quadraspidiotus pyri (Lichtenstein) (also known as pear scale)
pale booklouse	*Psyllipsocus ramburii* Sélys-Longchamps
	syn. *Nymphopsocus destructor* Enderlein
pale booklouse, larger	*Trogium pulsatorium* (L.)
pale brindled beauty moth	*Apocheima pilosaria* (Denis & Schiffermüller)
	syn. *Phigalia pedaria* (Fabricius)
	P. pilosaria (Denis & Schiffermüller)
pale mottled willow moth, *see* mottled willow moth, pale	
pale spotted gooseberry sawfly, *see* gooseberry sawfly, pale spotted	
pale tussock moth	*Calliteara pudibunda* (L.)
larva = hop-dog	syn. *Dasychira pudibunda* (L.)
palm millepede	*Choneiulus palmatus* (Nemec)
*palm thrips	*Thrips palmi* Karny
	(*see also* palm thrips, banded wing)
palm thrips, banded-wing	*Parthenothrips dracaenae* (Heeger)
parallel-sided ground beetle	*Abax parallelepipedus* (Piller & Mitterpacher)
parasitic flies	TACHINIDAE
	syn. LARVAEVORIDAE
parent bug	*Elasmucha grisea* (L.)
parsnip flower midge	*Kiefferia pericarpiicola* (Bremi)
parsnip moth	*Depressaria pastinacella* (Duponchel)
larva = parsnip webworm	misident. *D. heracliana* (L.)
parsnip webworm, *larva of* parsnip moth	
pattern engraver beetles	{ *Ips acuminatus* (Gyllenhal)
	Orthotomicus laricis (Fabricius)
pea and bean weevil	*Sitona lineatus* (L.)

COMMON NAMES—Arthropods
Common Name   Scientific Name

pea aphid   *Acyrthosiphon pisum* (Harris)
                  syn. *A. pisi* (Kaltenbach)
                      A. onobrychis (Boyer de Fonscolombe)
                      Macrosiphum onobrychis (Boyer de Fonscolombe)
                      M. pisi (Kaltenbach)

pea beetle   *Bruchus pisorum* (L.)
pea beetles, bean beetles   *Bruchus* spp.
peach aphid   *Brachycaudus schwartzi* (Börner)
              syn. *B. amygdali* (Buckton)
                 Anuraphis amygdali (Buckton)
                 Appelia schwartzi (Börner)

peach—potato aphid   *Myzus persicae* (Sulzer)
†peach—potato aphid parasitic wasps   { *Aphidius matricariae* Haliday
                 Aphidius picipes (Nees)
                 syn. *A. avenae* Haliday

peach scale   *Parthenolecanium persicae* (Fabricius)
              syn. *lecanium persicae* (Fabricius)

*peach scale, white   *Pseudaulacaspis pentagona* (Targioni-Tozzetti)
*peach twig borer   (moth) larva of *Anarsia lineatella* Zeller
pea leaf miners   (fly) larvae of   *Liriomyza congesta* (Becker)
                          Liriomyza pisivora Hering

pea midge   *Contarinia pisi* (Winnertz)
    pea mite, *see* red-legged earth mite
pea moth   *Cydia nigricana* (Fabricius)
              syn. *Laspeyresia nigricana* (Fabricius)

    pear and cherry sawfly, *see* pear slug sawfly
pear and cherry slugworm   (sawfly) larva of *Caliroa cerasi* (L.)
              syn. *C. limacina* (Retzius)

pear—bedstraw aphid   *Dysaphis pyri* (Boyer de Fonscolombe)
              syn. *Anuraphis pyri* (Boyer de Fonscolombe)
                 Sappaphis pyri (Boyer de Fonscolombe)

    pear bryobia (mite), *see* grass—pear bryobia
pear—coltsfoot aphid   *Anuraphis farfarae* (Koch)
pear—grass aphid   *Melanaphis pyraria* (Passerini)
              syn. *Geoktapia pyraria* (Passerini)
                 Longiunguis pyrarius (Passerini)

pear leaf blister mite   *Eriophyes pyri* (Pagenstecher)
              syn. *Phytoptus pyri* Pagenstecher
              misspelling *Phytoptus piri* Pagenstecher

pear leaf blister moth   *Leucoptera malifoliella* (Costa)
              syn. *Leucoptera scitella* (Zeller)

    pear leaf-curling midge, *see* pear leaf midge
pear leaf midge   *Dasineura pyri* (Bouché)
†pearly green lacewing   *Chrysopa perla* (L.)
pearly underwing moth   *Peridroma saucia* (Hübner)
    larva = variegated cutworm   syn. *Agrotis saucia* (Hübner)
                 Peridroma margaritosa (Haworth)
              misident. *P. porphyrea* (Denis & Schiffermüller)

pear midge   *Contarinia pyrivora* (Riley)
pear—parsnip aphid   *Anuraphis subterranea* (Walker)
    pear psylla, *see* pear sucker
    pear psyllid, *see* pear sucker
pear rust mite   *Epitrimerus piri* (Nalepa)
              misspelling *Epitrimerus pyri* (Nalepa)

COMMON NAMES—Arthropods

Common Name	Scientific Name
pear sawfly	*Hoplocampa brevis* (Klug)
pear sawfly, social	*Neurotoma saltuum* (L.)
	syn. *N. flaviventris* (Retzius)
pear scale, *see* oystershell scales	
pear slug sawfly	*Caliroa cerasi* (L.)
larva = pear and cherry slugworm	syn. *C. limacina* (Retzius)
pear sucker (psyllid)	*Psylla pyricola* Förster
	syn. *P. simulans* Förster
pear thrips	*Taeniothrips inconsequens* (Uzel)
	syn. *T. pyri* (Daniel)
pear weevil	*Magdalis barbicornis* (Latreille)
pea seed beetle, *see* pea beetle	
pea thrips	*Kakothrips pisivorus* (Westwood)
	syn. *Frankliniella robusta* (Uzel)
	Kakothrips pisivora (Westwood)
	K. robustus (Uzel)
pea weevils, bean weevils, clover weevils	*Sitona* spp.
pelargonium aphid	*Acyrthosiphon malvae* (Mosley)
	syn. *A. pelargonii* (Kaltenbach)
	Macrosiphum pelargonii
	(Kaltenbach)
pepper and salt moth, *see* peppered moth	
peppered moth	*Biston betularia* (L.)
larva = hop-cat	syn. *Amphidasis betularia* (L.)
permanent apple aphid, *see* green apple aphid	
permanent blackberry aphid, *see* blackberry aphid, permanent	
permanent carrot aphid, *see* carrot aphid, permanent	
permanent currant aphid, *see* currant aphid, permanent	
permanent dock aphid, *see* dock aphid, permanent	
pernicious scale, *see* San José scale	
Persian walnut leaf blister mite, *see* walnut leaf gall mite	
Peruvian larder beetle	*Dermestes peruvianus* Laporte de Castelnau
Pharaoh's ant	*Monomorium pharaonis* (L.)
phillyrea whitefly	*Siphoninus phillyreae* (Haliday)
phoenix (moth)	*Eulithis prunata* (L.)
*phoracantha borer	(beetle) larva of *Phoracantha semipunctata* (Fabricius)
phorid flies, *see* scuttle flies	
phylloxeras	PHYLLOXERIDAE
phylloxeras, aphids, hoppers, mealybugs, psyllids, scale insects, whiteflies	HOMOPTERA
pigeon body louse	*Hohorstiella lata* (Piaget)
pigeon bug	*Cimex columbarius* Jenyns
pigeon soft tick	*Argas reflexus* (Fabricius)
pigeon wing louse	*Columbicola columbae* (L.)
	syn. *C. bacula* (Nitzsch)
	C. filiformis (von Olfers)
pig follicle mite	*Demodex phylloides* Csokor
pig head mange mite, *see* pig follicle mite	
pig itch mite, *see* itch mite	
pig louse	*Haematopinus suis* (L.)
pillbug, common (woodlouse)	*Armadillidium vulgare* (Latreille)
pill millepede	*Glomeris marginata* (Villers)
pine aphid, large	*Cinara pinea* (Mordvilko)
pine aphid, narrow brown	*Eulachnus rileyi* (Williams)
	syn. *Protolachnus bluncki* (Börner)
pine aphid, narrow green	*Eulachnus brevipilosus* (Börner)
*pineapple scale	*Diaspis bromeliae* (Kerner)
pine bark beetles	{ *Hylurgops palliatus* (Gyllenhal)
	Pityogenes chalcographus (L.)

COMMON NAMES—Arthropods

Common Name	Scientific Name
pine bark beetle, six-toothed	*Ips sexdentatus* (Börner)
pine beauty moth	*Panolis flammea* (Denis & Schiffermüller)
	syn. *P. piniperda* (Panzer)
pine beetle, two-toothed	*Pityogenes bidentatus* (Herbst)
pine bud moth	*Blastesthia turionella* (L.)
	syn. *Evetria turionana* (Haworth)
pine carpet moth, grey	*Thera obeliscata* (Hübner)
pine carpet moth, red	*Thera firmata* (Hübner)
pine cone moth	*Cydia conicolana* (Heylaerts)
	syn. *Laspeyresia conicolana* (Heylaerts)
pine cone weevil	*Pissodes validirostris* (Sahlberg)
pine flat bug	*Aradus cinnamomeus* (Panzer)
pine gall mite, see pine twig-knot mite	
pine knothorn moth	*Dioryctria abietella* (Denis & Schiffermüller)
†pine ladybird (beetle)	*Anatis ocellata* (L.)
pine longhorn beetle	*Asemum striatum* (L.)
pine looper	(moth) larva of *Bupalus piniaria* (L.)
†pine looper parasites (wasps)	*Cratichneumon viator* (Scopoli)
	syn. *C. nigritarius* (Gravenhorst)
	Dusona oxyacanthae (Boie)
	syn. *Campoplex oxyacanthae* Boie
	Heteropelma calcator (Wesmael)
	Polytribax arrogans (Gravenhorst)
	syn. *Plectocryptus arrogans* (Gravenhorst)
pine needle gall midge	*Contarinia baeri* Prell
	syn. *Cecidomyia baeri* (Prell)
pine needle scale	*Matsucoccus pini* (Green)
pine needle-shortening gall midge	*Thecodiplosis brachyntera* (Schwägrichen)
pine needle weevil	*Brachonyx pineti* (Paykull)
pine resin-gall moth	*Petrova resinella* (L.)
	syn. *Evetria resinella* (L.)
pine resin moth	*Cydia coniferana* (Ratzeburg)
	syn. *Laspeyresia coniferana* (Ratzeburg)
pine root aphid	*Stagona pini* (Burmeister)
	syn. *Prociphilus pini* (Burmeister)
pine sawfly	*Diprion pini* (L.)
	syn. *Lophyrus pini* (L.)
pine sawfly, large, see pine sawfly	
pine sawfly, lesser } see fox-coloured sawfly	
pine sawfly, small }	
pine shoot beetle	*Tomicus piniperda* (L.)
	syn. *Blastophagus piniperda* (L.)
	Myelophilus piniperda (L.)
pine shoot beetle, lesser	*Tomicus minor* (Hartig)
pine shoot moth	*Rhyacionia buoliana* (Denis & Schiffermüller)
	syn. *Evetria buoliana* (Denis & Schiffermüller)
pine tortrix (moth)	*Epinotia nanana* (Treitschke)
pine-twig bark beetle	*Pityophthorus pubescens* (Marsham)
pine twig gall mite, see pine twig-knot mite	
pine twig-knot mite	*Trisetacus pini* (Nalepa)
pine weevil	*Hylobius abietis* (L.)
pin-hole borers, shot-hole borers	(beetle) larvae of PLATYPODIDAE (in particular *Platypus* spp.)
	Xyleborus spp.
*pink scavenger	(moth) larva of *Pyroderces rileyi* (Walsingham)
	syn. *Sathrobrota rileyi* (Walsingham)

COMMON NAMES—Arthropods

Common Name	Scientific Name
pistol casebearer	(moth) larva of *Coleophora anatipennella* (Hübner)
	syn. *Eupista anatipennella* (Hübner)
pith moth	*Spuleria atra* (Haworth)
	syn. *Blastodacna atra* (Haworth)

place bug, *see* forest bug
plain yellow twist (moth), *see* timothy tortrix

plantain aphid	*Aphis plantaginis* Goeze

plantation fly, *see* sheep head fly
plant-lice, *see* aphids

plaster beetles	LATHRIDIIDAE

plum cambium midge, *see* plum cambium miner

plum cambium miner	(fly) larva of *Phytobia cerasiferae* (Kangas)
	syn. *Dendromyza cerasiferae* Kangas
plum curculio (weevil)	*Conotrachelus nenuphar* (Herbst)
plum fruit moth	*Cydia funebrana* (Treitschke)
larva = red plum maggot	syn. *Ernarmonia funebrana* (Treitschke)
	Grapholita funebrana (Treitschke)
	Laspeyresia funebrana (Treitschke)

plum gall mite, *see* big-beaked plum mite

plum leaf gall mite	*Eriophyes padi* (Nalepa)
	syn. *Phytoptus padi* (Nalepa)

plum leaf mite, *see* plum rust mite *and* plum spur mite

plum leaf sawfly	*Priophorus pallipes* (Lepeletier)
	syn. *P. varipes* (Lepeletier)
	misident. *P. padi* (L.)
plum pouch-gall mite	*Eriophyes similis* (Nalepa)
	syn. *Phytoptus similis* Nalepa
plum rust mite	*Aculus fockeui* (Nalepa & Trouessart)
	syn. *Phyllocoptes fockeui* Nalepa & Trouessart
	Phytoptus fockeui (Nalepa & Trouessart)
	Vasates fockeui (Nalepa & Trouessart)
plum sawfly	*Hoplocampa flava* (L.)
plum spur mite,	*Acalitus phloeocoptes* (Nalepa)
	syn. *Aceria phloeocoptes* (Nalepa)
	Phytoptus phloeocoptes Nalepa
plum tortrix (moth)	*Hedya pruniana* (Hübner)
	syn. *Apotomis pruniana* (Hübner)
	Argyroploce pruniana (Hübner)

plum weevil, *see* red-legged weevil

plump white springtail	*Anurida granaria* (Nicolet)
plusia moths	PLUSIINAE
larvae = semi-loopers	

pod midge, *see* brassica pod midge
pollen beetles, *see* blossom beetles

polygonum leaf beetle	*Gastrophysa polygoni* (L.)
polygonum gall midge	*Wachtliella persicariae* (L.)
poplar and willow borer	(beetle) larva of *Saperda carcharias* (L.)
poplar and willow cambium miner	(fly) larva of *Phytobia cambii* (Hendel)
	syn. *Dizygomyza barnesi* Hendel
	D. carbonaria (Zetterstedt)
poplar borer, small	(beetle) larva of *Saperda populnea* (L.)
poplar—buttercup aphid	*Thecabius affinis* (Kaltenbach)
	syn. *Pemphigus affinis* (Kaltenbach)
poplar cloaked bell moth	*Gypsonoma aceriana* (Duponchel)
poplar—cudweed aphid	*Pemphigus populinigrae* (Schrank)
	syn. *P. filaginis* (Boyer de Fonscolombe)

COMMON NAMES—Arthropods

Common Name	Scientific Name
poplar flea beetle	*Chalcoides aurea* (Fourcroy) misident. *Crepidodera helxines* (L.)
poplar gall midge	*Contarinia petioli* (Kieffer) syn. *Syndiplosis petioli* (Kieffer)
poplar hawk moth	*Laothoe populi* (L.)
poplar leaf aphid	*Chaitophorus leucomelas* Koch
poplar leaf beetle, red	*Chrysomela populi* (L.) syn. *Melasoma populi* (L.)
poplar leaf beetle, small	*Phyllodecta laticollis* Suffrian syn. *P. cavifrons* Thomson
poplar leaf gall aphid, *see* poplar—buttercup aphid	
poplar leaf gall midge	*Harmandia loewi* (Rübsaamen)
poplar leaf miner	(fly) larva of *Paraphytomyza populi* (Kaltenbach) syn. *Napomyza populi* (Kaltenbach) *Phytagromyza populi* (Kaltenbach)
poplar leaf miner	(moth) larva of *Phyllocnistis unipunctella* (Stephens)
poplar leaf roller weevil	*Byctiscus populi* (L.)
poplar—lettuce aphid, *see* lettuce root aphid	
poplar longhorn beetle, large *larva* = poplar and willow borer	*Saperda carcharias* (L.)
poplar longhorn beetle, small *larva* = small poplar borer	*Saperda populnea* (L.)
poplar sawflies	*Pristiphora conjugata* (Dahlbom) *Trichiocampus viminalis* (Fallén) syn. *Cladius viminalis* (Fallén) *Priophorus viminalis* (Fallén)
poplar sawfly, gregarious	*Nematus melanaspis* Hartig syn. *Pteronidea melanaspis* (Hartig)
poplar shoot aphid	*Chaitophorus populeti* (Panzer)
poplar spiral-gall aphid	*Pemphigus spyrothecae* Passerini misspelling *P. spirothecae* Passerini
poppy gall cynipids (wasps)	*Aylax minor* Hartig *Aylax papaveris* (Perris)
poppy leaf roller	(moth) larva of *Cnephasia asseclana* (Denis & Schiffermüller)
porphyry knot—horn moth	*Numonia suavella* (Zincken) syn. *Eurhodope suavella* (Zinchen)
post boring beetle	*Ptilinus pectinicornis* (L.)
potato aphid	*Macrosiphum euphorbiae* (Thomas) syn. *M. solanifolii* Ashmead misident. *M. gei* (Koch)
†potato aphid hover fly	*Platycheirus manicatus* (Meigen)
potato capsid (bug)	*Calocoris norvegicus* (Gmelin)
potato flea beetle	*Psylliodes affinis* (Paykull)
potato ground bug	*Peritrechus lundi* (Gmelin)
potato leafhoppers	*Eupterycyba jucunda* (Herrich-Schäffer) syn. *Typhlocyba jucunda* Herrich-Schäffer *Eupteryx aurata* (L.) syn. *Cicadella aurata* (L.)
*potato moth	*Phthorimaea operculella* (Zeller) syn. *Gnorimoschema operculella* (Zeller)
potato sawfly	*Pachyprotasis variegata* (Fallén)
potato scab gnat	*Pnyxia scabiei* (Hopkins)
potato skin borer, *see* potato stem borer	
potato stem borer	(moth) larva of *Hydraecia micacea* (Esper) syn. *Gortyna micacea* (Esper)
potato thrips, *see* onion thrips	
potato tuber moth, *see* potato moth	

COMMON NAMES—Arthropods

Common Name	Scientific Name
potentilla gall cynapids (wasps)	*Xestophanes* spp.
poultry-house fly	*Ophyra ignava* (Harris)
	syn. *O. leucostoma* (Wiedemann)
poultry litter mite	*Androlaelaps casalis* (Berlese)
	syn. *Haemolaelaps casalis* (Berlese)
poultry red mite, *see* chicken mite	
powdered quaker moth	*Orthosia gracilis* (Denis & Schiffermüller)
power-post beetles	LYCTIDAE
	(in particular *Lyctus* spp. and *Minthea* spp.)
†powdery lacewings	{ *Coniopteryx* spp.
	Conwentzia spp.
	Semidalis aleyrodiformis (Stephens)
	misspelling *Semiadalis aleyrodiformis* (Stephens)
†predatory rove beetle, minute	*Oligota flavicornis* (Boisduval & Lacordaire)
privet aphid	*Myzus ligustri* (Mosley)
	syn. *Myzodes ligustri* (Kaltenbach)
privet thrips	*Dendrothrips ornatus* (Jablonowski)
prunus cambium miner, *see* plum cambium miner	
pseudoscorpions, *see* false scorpions	
psocids	PSOCOPTERA
psoroptic mange mite	*Psoroptes equi* (Raspail)
	syn. *P. bovis* (Hertwig)
	P. cuniculi (Delafond)
	P. ovis (Hering)
psyllids, aphids, hoppers, mealybugs, phylloxeras, scale insects, whiteflies	HOMOPTERA
psyllids, jumping plant-lice, suckers	PSYLLIDAE
pubic louse, *see* crab louse	
pulse beetles	BRUCHIDAE
puss moth	*Cerura vinula* (L.)
	syn. *Dicranura vinula* (L.)
pygmy beetle, *see* pygmy mangold beetle	
pygmy mangold beetle	*Atomaria linearis* Stephens
quill mite (of poultry)	*Syringophilus bipectinatus* Heller
rabbit ear canker mite, *see* psoroptic mange mite	
rabbit ear mite, *see* psoroptic mange mite	
rabbit flea, *see* European rabbit flea	
rabbit follicle mite	*Demodex cuniculi* (Pfeiffer)
rabbit fur mites	{ *Cheyletiella parasitivorax* (Mégnin)
	Listrophorus gibbus Pagenstecher
rabbit mange mite, *see* rabbit follicle mite	
rabbit tick	*Ixodes ventalloi* Gil Collado
rain beetle	*Pterostichus cuprea* (L.)
	syn. *Feronia cupreus* (L.)
raisin moth	*Ephestia figulilella* Gregson
rape winter stem weevil	*Ceutorhynchus picitarsis* Gyllenhal
raspberry aphid, *see* raspberry aphid, small	
raspberry aphid, large	*Amphorophora idaei* (Börner)
raspberry aphid, small	*Aphis idaei* van der Goot
raspberry beetle	*Byturus tomentosus* (Degeer)
raspberry borer	(moth) larva of *Lampronia rubiella* (Bjerkander)

COMMON NAMES—Arthropods

Common Name	Scientific Name
†raspberry bug	*Orius vicinus* (Ribaut) misident. *O. minutus* (L.)
raspberry cane midge	*Resseliella theobaldi* (Barnes) syn. *Thomasiniana theobaldi* Barnes
raspberry flea beetles	{ *Batophila aerata* (Marsham) { *Batophila rubi* (Paykull)
raspberry leaf and bud mite	*Phyllocoptes gracilis* (Nalepa) syn. *Aceria gracilis* (Nalepa) *Eriophyes gracilis* (Nalepa) *Phytoptus gracilis* (Nalepa)
raspberry leaf mining sawflies	{ *Metallus albipes* (Cameron) { syn. *Fenusa albipes* Cameron { *Metallus pumilus* (Klug)

raspberry maggot (moth larva), see raspberry borer
raspberry mite, see raspberry leaf and bud mite

raspberry moth	*Lampronia rubiella* (Bjerkander)
larva = raspberry borer	
raspberry sawfly	*Empria tridens* (Konow)
raspberry sawfly, small	*Priophorus morio* (Lepeletier) syn. *P. brullei* Dahlbom *P. tener* (Zaddach) incorrect attribution to author *P. tener* (Hartig)

raspberry stem gall midge, see blackberry stem gall midge

rat ear mange mite	*Notoedres muris* (Mégnin)
rat flea	*Nosopsyllus fasciatus* (Bosc)
rat fur mite	*Radfordia ensifera* (Poppe)
rat head itch mite } see rat ear mange mite	
rat head mange mite }	
rat louse	*Polyplax spinulosa* (Burmeister)
rat myobiid mite, see rat fur mite	
rat tailed maggots	(fly) larvae of { *Eristalis pertinax* (Scopoli) { *E. tenax* (L.)
red and black froghopper	*Cercopis vulnerata* Illiger in Rossi syn. *C. sanguinea* (Geoffroy in Fourcroy)
red ants	{ *Myrmica rubra* (L.) { *Myrmica ruginodis* Nylander

†red apple capsid (bug), see apple capsid, red
red barred tortrix (moth), see fruit tree tortrices
red-belted clearwing, see apple clearwing moth

red bud borer	(fly) larva of *Resseliella oculiperda* (Rübsaamen) syn. *Thomasiniana oculiperda* (Rübsaamen)
red clover gall gnat	*Campylomyza ormerodi* (Kieffer)
red clover seed weevils, see clover seed weevils	
red currant—arrowgrass aphid	*Aphis triglochinis* Theobald
red currant blister aphid	*Cryptomyzus ribis* (L.) syn. *Capitophorus ribis* (L.)
*red-headed ash borer	(beetle) larva of *Neoclytus acuminatus* (Fabricius)
red ichneumons (wasps)	*Ophion* spp.
red-legged earth mite	*Penthaleus major* (Dugès)
red-legged ham beetle, see copra beetle	
†red-legged ichneumon (wasp)	*Pimpla hypochondriaca* (Retzius) syn. *P. instigator* (Fabricius)
red-legged weevil	*Otiorhynchus clavipes* (Bonsdorff)
red oak borer	(beetle) larva of *Enaphalodes rufulum* (Haldeman) syn. *Romaleum rufulum* Haldeman

COMMON NAMES—Arthropods

Common Name   **Scientific Name**

red pepper mites	PYGMEPHORIDAE (in particular *Pygmephorus* spp.)
red pine carpet moth, *see* pine carpet moth, red	
red plum maggot	(moth) larva of *Cydia funebrana* (Treitschke) syn. *Ernarmonia funebrana* (Treitschke) *Grapholita funebrana* (Treitschke) *Laspeyresia funebrana* (Treitschke)
red poplar leaf beetle, *see* poplar leaf beetle, red	
red-shouldered ham beetle	*Necrobia ruficollis* (Fabricius)
*red-shouldered powder-post beetle	*Xylobiops basilaris* (Say)
red spider mite, *see* two-spotted spider mite	
red spider mite predator (mite), *see* two-spotted spider mite predator	
red sword-grass moth, *see* sword-grass moth, red	
†red velvet mite	*Allothrombium fuliginosum* (Hermann)
red wasp	*Vespula rufa* (L.)
reed gall fly	*Lipara lucens* Meigen
rhododendron bug	*Stephanitis rhododendri* Horváth
rhododendron hopper	*Graphocephala fennahi* Young syn. *G. coccinea* (Förster)
rhododendron whitefly	*Dialeurodes chittendeni* Laing
rhododendron whitefly, small, *see* azalea whitefly	
rhomboid tortrix (moth), *see* fruit tree tortrices	
rhynchites (weevils)	*Rhynchites* spp.
rice moth	*Corcyra cephalonica* (Stainton)
rice weevil, greater, *see* maize weevil	
rice weevil	*Sitophilus oryzae* (L.)
rice weevil, lesser, *see* rice weevil	
robber flies	ASILIDAE
robin's pincushion	gall induced by (wasp) larva of *Diplolepis rosae* (L.) syn. *Rhodites rosae* (L.)
root mealybugs	*Rhizoecus* spp.
rose aphid	*Macrosiphum rosae* (L.)
rose aphid, lesser	*Myzaphis rosarum* (Kaltenbach)
rose chafer (beetle)	*Cetonia aurata* (L.)
rose—grain aphid	*Metopolophium dirhodum* (Walker) syn. *Macrosiphum dirhodum* (Walker)
rose-hip fly	*Rhagoletis alternata* (Fallén) syn. *Spilographa alternata* (Fallén) *Zonosema alternata* (Fallén)
rose leafhopper	*Edwardsiana rosae* (L.) syn. *Typhlocyba rosae* (L.)
rose leaf midge	*Dasineura rosarum* (Hardy) syn. *Wachtiiella rosarum* (Hardy)
rose leaf miner	(moth) larva of *Stigmella anomalella* (Goeze) syn. *Nepticula anomalella* (Goeze) *Stigmella rosella* (Schrank)
rose maggots (moth larvae), *see* leaf rollers	
rose root aphid	*Maculolachnus submacula* (Walker) syn. *Lachnus rosae* Cholodkovsky *Pterochlorus rosae* (Cholodkovsky)
rose sawfly, *see* rose slug sawfly	
rose sawfly, large	*Arge ochropus* (Gmelin in Linnaeus)
rose scale	*Aulacaspis rosae* (Bouché) syn. *Diaspis rosae* Bouché
rose slug sawfly	*Endelomyia aethiops* (Fabricius) syn. *Caliroa aethiops* (Fabricius)
rose smooth pea-gall cynipid (wasp)	*Diplolepis eglanteriae* (Hartig)
rose spherical-gall cynipid (wasp)	*Diplolepis spinosissimae* (Giraud)

COMMON NAMES—Arthropods

Common Name	Scientific Name
rose-spiked pea-gall cynipid (wasp)	*Diplolepis nervosa* (Curtis) syn. *D. dispar* (Niblet) *Rhodites nervosus* (Curtis)
rose stem gall cynipids (wasps)	{ *Diastrophus rubi* (Bouché) { *Diplolepis mayri* (Schlechtendal)
rose thrips	*Thrips fuscipennis* Haliday syn. *T. menyanthidis* Bagnall *Thrips major* Uzel (also known as rubus thrips)
rose tortrix (moth)	*Archips rosana* (L.)
rose twist (moth), *see* rose tortrix	
rosy apple aphid	*Dysaphis plantaginea* (Passerini) syn. *Anuraphis roseus* (Baker) *Sappaphis mali* (Ferrari) *S. plantaginea* (Passerini) misident. *Ceruraphis malifoliae* (Fitch)
rosy leaf-curling aphid	*Dysaphis devecta* (Walker) syn. *Sappaphis devecta* (Walker) misident. *Anuraphis crataegi* (Kaltenbach) (in part)
rosy minor moth	*Mesoligia literosa* (Haworth) syn. *Miana literosa* (Haworth) *Procus literosa* (Haworth)
rosy rustic moth	*Hydraecia micacea* (Esper)
larva = potato stem borer	syn. *Gortyna micacea* (Esper)
rove beetles	STAPHYLINIDAE
rove beetles, small	*Atheta* spp.
†rove beetles, small-headed	*Aleochara* spp.
rubus aphid, *see* bramble aphid	
rubus leafhoppers	{ *Macropsis fuscula* (Zetterstedt) { *Macropsis scotti* Edwards
rubus thrips, *see* rose thrips	
rush sucker (psyllid)	*Livia juncorum* (Latreille)
*rustic borer	(beetle) larva of *Xylotrechus colonus* (Fabricius)
rustic moth, common	*Mesapamea secalis* (L.) syn. *Apamea secalis* (L.) *Celaena secalis* (L.)
rustic shoulder knot moth	*Apamea sordens* (Hufnagel)
larva = wheat cutworm	syn. *A. basilinea* (Denis & Schiffermüller)
rustic tortrix (moth)	*Clepsis senecionana* (Hübner)
rustic twist (moth), *see* rustic tortrix	
rust-red flour beetle	*Tribolium castaneum* (Herbst)
rust-red grain beetle	*Cryptolestes ferrugineus* (Stephens) syn. *Laemophloeus ferrugineus* (Stephens)
*rust-winged longhorn beetle	*Tetropium cinnamopterum* Kirby
rusty longhorn beetle	*Arhopalus rusticus* (L.) syn. *Criocephalus rusticus* (L.)
rusty tussock, *see* vapourer moth	
saddle gall midge	*Haplodiplosis marginata* (von Roser) syn. *H. equestris* (Wagner)
sail wasps, *see* ichneumons	
sainfoin flower midge	*Contarinia onobrychidis* Kieffer
sainfoin leaf midge	*Bremiola onobrychidis* (Bremi)
Saint, *see* St.	
sallow flat-body moth, *see* willow shoot moth	

COMMON NAMES—Arthropods

Common Name	Scientific Name
sallow stem galler (fly)	*Hexomyza schineri* (Giraud)
salt marsh mosquitoes	*Aedes caspius* (Pallas)
	Aedes detritus (Haliday)
sand martin tick	*Ixodes lividus* Koch
sand weevil	*Philopedon plagiatus* (Schaller)
	syn. *Cneorhinus plagiatus* (Schaller)
*San José scale	*Comstockaspis perniciosa* (Comstock)
	syn. *Quadraspidiotus perniciosus* (Comstock)
	Aspidiotus perniciosus Comstock

    sarcoptic mange mite (of goat), *see* itch mite

satin moth	*Leucoma salicis* (L.)
	syn. *Stilpnotia salicis* (L.)
sawflies, ants, bees, wasps, etc.	HYMENOPTERA
saw-toothed grain beetle	*Oryzaephilus surinamensis* (L.)
*sawyer beetles	*Monochamus* spp.

    scabies mite, *see* itch mite

scab mites	PSOROPTIDAE

    (*see also* dust mites and bird ked mites)
    scale aphid, *see* orchid aphid

†scale insect parasite (wasp)	*Aphytis mytilaspidis* (LeBaron)
scale insects	{ COCCIDAE (also known as soft scales)
	DIASPIDIDAE (also known as armoured scales)
	MARGARODIDAE }
scale insects, aphids, hoppers, mealybugs, phylloxeras, psyllids, whiteflies	HOMOPTERA
scalloped hazel moth	*Odontopera bidentata* (Clerck)
	syn. *Gonodontis bidentata* (Clerck)
scaly grain mite	*Suidasia nesbitti* Hughes
scalyleg mite (of budgerigars)	*Knemidokoptes pilae* Lavoipierre & Griffiths
scalyleg mite (of fowl)	*Knemidokoptes mutans* (Robin & Lanquetin)

    scaly strawberry weevil, *see* strawberry weevil, scaly

Scandinavian spruce sawfly	*Pristiphora subartica* (Forsslund)

    scarce blackberry aphid, *see* blackberry aphid, scarce
    scarce bordered straw moth, *see* Old World bollworm

scarce umber moth	*Agriopis aurantiaria* (Hübner)
	syn. *Erannis aurantiaria* (Hübner)
	Hybernia aurantiaria (Hübner)
*scarlet-coated longhorn beetle	*Pyrrhidium sanguineum* (L.)

    Schenck's gall wasp, *see* oak leaf smooth-gall cynipid (wasp)

sciarid flies	*Bradysia* spp.
	misident. *Sciara* spp.
scorpion-flies	MECOPTERA
scorpion-fly, common	*Panorpa communis* L.
Scots pine adelges	*Pineus pini* (Macquart)
	incorrect attribution to author *P. pini* (Gmelin in Linnaeus)
*screw worm	(fly) larva of *Cochliomyia hominivorax* (Coquerel)

    scurfy scale, *see* rose scale

scuttle flies	PHORIDAE
seabird tick	*Ixodes uriae* White
	syn. *Ixodes putus* (Cambridge)
seaweed flies	{ *Coelopa frigida* (Fabricius)
	Coelopa pilipes Haliday }

    seed mite, *see* grainstack mite

seed wasps	*Megastigmus* spp.
semi-loopers	(moth) larvae of PLUSIINAE

COMMON NAMES—Arthropods

Common Name	Scientific Name
setaceous Hebrew character moth	*Xestia c-nigrum* (L.)
larva = spotted cutworm	syn. *Amathes c-nigrum* (L.)
†seven-spot ladybird (beetle)	*Coccinella septempunctata* L.
	syn. *C. 7-punctata* L.
sewage farm flies, *see* moth flies	
sewage filter-bed flies, window gnats	ANISOPODIDAE
	(in particular *Sylvicola* spp.)
sewage fly	*Leptocera caenosa* Rondani
sewage mites	*Histiostoma* spp.
sexton beetles, *see* burying beetles	
shallot aphid	*Myzus ascalonicus* Doncaster
shasta daisy midge	*Contarinia chrysanthemi* (Kieffer)
sheep biting louse	*Bovicola ovis* (Schrank)
	syn. *Damalina ovis* (Schrank)
	D. ovisarietis (Schrank)
	D. sphaerocephala (von Olfers)
sheep blow-fly, *see* sheep maggot fly	
sheep bot fly, *see* sheep nostril fly	
sheep chorioptic mange mite, *see* chorioptic mange mite	
sheep follicle mite	*Demodex ovis* Railliet
*sheep foot louse	*Linognathus pedalis* (Osborn)
sheep head fly	*Hydrotaea irritans* (Fallén)
sheep itch mite, *see* itch mite	
sheep ked (fly)	*Melophagus ovinus* (L.)
sheep maggot fly	*Lucilia sericata* (Meigen)
sheep mange mite, *see* sheep follicle mite	
sheep nostril fly	*Oestrus ovis* L.
sheep scab mite, *see* psoroptic mange mite	
sheep's sorrel gall weevil	*Apion haematodes* Kirby
	syn. *A. frumentarium* (Paykull)
sheep sucking louse	*Linognathus ovillus* (Neumann)
sheep tick	*Ixodes ricinus* (L.)
shield bugs	ACANTHOSOMIDAE
	CYDNIDAE
	PENTATOMIDAE
	SCUTELLERIDAE
shiner, *see* German cockroach	
shining smut beetles	*Stilbus* spp.
shiny spider beetle	*Gibbium psylloides* (de Czenpinski)
ship cockroach, *see* American cockroach	
ship timberworm	(beetle) larva of *Lymexylon navale* (L.)
shore bugs, jumping bugs	SALDIDAE
shore flies	EPHYDRIDAE
short-horned grasshoppers, *see* grasshoppers and locusts	
short-nosed cattle louse	*Haematopinus eurysternus* (Nitzsch)
shot-hole borers, pin-hole borers	(beetle) larvae of PLATYPODIDAE
	(in particular *Platypus* spp.)
	Xyleborus spp.
shy bug, *see* hop capsid	
*Siamese grain beetle	*Lophocateres pusillus* (Klug)
†*silk moth, common	*Bombyx mori* (L.)
larva = mulberry silkworm	
silkworm (moth *larva*), *see* mulberry silkworm	
silver birch aphid	*Euceraphis betulae* (Koch)
silver fir adelges	*Adelges piceae* (Ratzeburg)
	syn. *A. piceae canadensis* Merker & Eichhorn
	Dreyfusia piceae (Ratzeburg)

COMMON NAMES—Arthropods

Common Name	Scientific Name
silver fir bark beetle	*Dryocoetes autographus* (Ratzeburg)
silver fir migratory adelges	*Adelges nordmannianae* (Eckstein)
	syn. *A. nuesslini* (Börner)
	A. schneideri (Börner)
	Dreyfusia nordmannianae (Eckstein)
*silver fir seed wasp	*Megastigmus pinus* Parfitt
silverfish (bristletail)	*Lepisma saccharina* L.
silver-green leaf weevil	*Phyllobius argentatus* (L.)

silver top (of cereals and grasses), *see* grass and cereal mite

silver y moth	*Autographa gamma* (L.)
	syn. *Plusia gamma* (L.)
†silver y moth parasite (fly)	*Phryxe vulgaris* (Fallén)

six-toothed pine bark beetle, *see* pine bark beetle, six-toothed

skin moth	*Monopis rusticella* (Hübner)

skipjacks, *see* click beetles

slender burnished brass moth	*Diachrysia orichalcea* (Fabricius)
	syn. *Plusia orichalcea* (Fabricius)

slender duck louse, *see* duck wing louse

slender grey capsid (bug)	*Dicyphus errans* (Wolff)
slender-horned flour beetle	*Gnatocerus maxillosus* (Fabricius)

slender pigeon louse, *see* pigeon wing louse

sloe bug	*Dolycoris baccarum* (L.)
slug- and snail-killing flies	SCIOMYZIDAE
slug mite	*Riccardoella limacum* (Schrank)

small banded pine weevil, *see* banded pine weevil, small
small black flea beetle, *see* black flea beetle, small
small blue cattle louse, *see* blue cattle louse
small broom bark beetle, *see* broom bark beetle, small
small chrysanthemum aphid, *see* chrysanthemum aphid, small

small eggar moth	*Eriogaster lanestris* (L.)

small elm bark beetle, *see* elm bark beetle, small
smaller eight-toothed spruce bark beetle, *see* spruce bark beetle, smaller eight-toothed
small ermine moths, *see* ermine moths, small

*small-eyed flour beetle	*Palorus ratzeburgi* (Wissmann)

small fruit flies, *see* fruit flies, small
small gooseberry sawfly, *see* gooseberry sawfly, small
small-headed rove beetles, *see* rove beetles, small-headed
small horse fly, *see* horse fly, small
small larch sawfly, *see* larch sawfly, small
small mottled willow moth, *see* mottled willow moth, small
small narcissus flies, *see* narcissus flies, small
small pine sawfly, *see* fox-coloured sawfly
small poplar borer, *see* poplar borer, small
small poplar leaf beetle, *see* poplar leaf beetle, small
small poplar longhorn beetle, *see* poplar longhorn beetle, small
small raspberry aphid, *see* raspberry aphid, small
small raspberry sawfly, *see* raspberry sawfly, small
small rhododendron whitefly, *see* azalea whitefly
small rove beetles, *see* rove beetles, small
small spruce adelges, *see* spruce adelges, small
small spruce bark beetle, *see* spruce bark beetle, small
small spruce woolly aphid, *see* spruce adelges, small
small striped flea beetle, *see* striped flea beetle, small

small tabby moth	*Aglossa caprealis* (Hübner)

larva = murky meal caterpillar
small walnut aphid, *see* walnut aphid, small
small white butterfly, *see* white butterfly, small
small willow aphid, *see* willow aphid, small
small winter moth, *see* winter moth

COMMON NAMES—Arthropods

Common Name	Scientific Name
smilax thrips	*Hercinothrips bicinctus* (Bagnall)
smooth broad-nosed weevil, *see* broad-nosed weevil, smooth	
smooth spangle gall wasp, *see* oak leaf smooth-gall cynipid (wasp)	
snail- and slug-killing flies	SCIOMYZIDAE
snail-killing fly	*Hydromya dorsalis* (Fabricius)
snake-flies	RAPHIDIOPTERA
snake-fly, common	*Raphidia notata* Fabricius
snake millepedes	{ *Archiboreoiulus pallidus* (Brade-Birks)
(*see also* spotted snake millepede,	*Boreoiulus tenuis* (Bigler)
black millepede *and* white-legged black millepede)	
snake mite	*Ophionyssus natricis* (Gervais)
snipe flies	{ *Atherix* spp. RHAGIONIDAE *Symphoromyia* spp.
social pear sawfly, *see* pear sawfly, social	
soft scales	COCCIDAE (also known as scale insects)
(*see also* brown soft scale)	
soft ticks	ARGASIDAE
soldier flies	STRATIOMYIDAE
solitary bees, *see* andrenas	
Solomon's seal sawfly	*Phymatocera aterrima* (Klug) misspelling *Phytomatocera aterrima* (Klug)
*South American longhorn beetle	*Trachyderes hilaris* Bates
sowbugs, *see* woodlice	
Spanish 'fly', *see* blister beetle	
spectrum wood wasp	*Xeris-spectrum* (L.)
spider beetles	PTINIDAE
†spider mite predators (mites), fruit tree red	{ *Amblyseius finlandicus* (Oudemans) syn. *Typhlodromus finlandicus* (Oudemans) *Typhlodromus pyri* Scheuten
†spider mite predator (mite), two-spot	*Phytoseiulus persimilis* Athias-Henriot syn. *P. riegeli* Dosse
spider mites	TETRANYCHIDAE
spiders	ARANEAE
spiders, mites, ticks, harvestmen, etc.	ARACHNIDA
spiked pox beetle	*Sinodendron cylindricum* (L.)
spinach beet bug	*Piesma maculatum* (Laporte de Castelnau)
spinach stem fly	*Delia echinata* (Séguy) syn. and misspelling *Hylemyia echinata* (Séguy)
spindle ermine moth	*Yponomeuta cagnagella* (Hübner) syn. *Y. cognatella* Treitschke
spiny rat mite	*Laelaps echidninus* Berlese
spiraea sawfly	*Nematus spiraeae* Zaddach & Brischke syn. *Pteronidea spiraeae* (Zaddach & Brischke)
spotted apple budworm (moth *larva*), *see* apple budworm, spotted	
spotted crane flies	*Nephrotoma* spp.
spotted cutworm	(moth) larva of *Xestia c-nigrum* (L.)
spotted millepede, *see* spotted snake millepede	
spotted pine aphid	*Eulachnus agilis* (Kaltenbach)
spotted snake millepede	*Blaniulus guttulatus* (Fabricius) incorrect attribution to author: *B. guttulatus* (Bosc)
springtails	COLLEMBOLA
spruce adelges, small	*Pineus pineoides* (Cholodkovsky)
spruce aphid, *see* green spruce aphid	

COMMON NAMES—Arthropods

Common Name	Scientific Name
spruce bark aphid, *see* spruce stem aphid	
spruce bark beetle	*Ips typographus* (L.)
*spruce bark beetle, great	*Dendroctonus micans* (Kugelann)
*spruce bark beetle, northern	*Ips duplicatus* (Sahlberg)
spruce bark beetle, small	*Polygraphus poligraphus* (L.)
*spruce bark beetle, smaller eight-toothed	*Ips amitinus* Eichhoff
spruce bug	*Gastrodes abietum* Bergroth
spruce cone gall midge	*Kaltenbachiola strobi* (Winnertz)
spruce gall adelges	*Pineus similis* (Gillette)
spruce mite, *see* conifer spinning mite	
spruce needle miner, (moth) *larva of* spruce needle tortrix	
spruce needle tortrix (moth)	*Epinotia tedella* (Clerck)
larva = spruce needle miner	syn. *Eucosma tedella* (Clerck)
spruce needle weevil	*Polydrusus pilosus* Gredler

spruce pineapple-gall adelges	*Adelges abietis* (L.)
	  syn. *A. gallarumabietis* (Degeer)
	    Sacchiphantes abietis (L.)
	Adelges viridis (Ratzeburg)
	  syn. *A. laricis* (Hartig)
	    Sacchiphantes viridis (Ratzeburg)
	  misident. *Adelges geniculatus* (Ratzeburg)

spruce root aphid	*Pachypappa tremulae* (L.) syn. *Asiphum tremulae* (L.)
spruce sawfly, gregarious	*Pristiphora abietina* (Christ) syn. *P. pini* (Retzius)
spruce shoot aphid, *see* brown spruce aphid	
spruce spider mite, *see* conifer spinning mite	
spruce stem aphid	*Cinara piceae* (Panzer)
spruce tip sawfly	*Pristiphora ambigua* (Fallén) syn. *P. furvescens* (Cameron)
spruce tip tortrix (moth)	*Zeiraphera ratzeburgiana* (Ratzeburg)
spruce twig aphid	*Mindarus obliquus* (Cholodkovsky)
spruce whorl scale	*Physokermes piceae* (Schrank)
square-necked grain beetle	*Cathartus quadricollis* (Guérin-Méneville)
stable fly	*Stomoxys calcitrans* (L.)
stable tabby moth, *see* small tabby moth	
stack bug	*Lyctocoris campestris* (Fabricius)
stag beetle, lesser	*Dorcus parallelipipedus* (L.)
stag beetles	LUCANIDAE
stag beetle	*Lucanus cervus* (L.)
steamfly, *see* German cockroach	
steel-blue wood wasps	*Sirex cyaneus* Fabricius *Sirex juvencus* (L.) *Sirex noctilio* Fabricius
steel-blue sawfly	*Acantholyda erythrocephala* (L.)
stem sawflies	CEPHIDAE
stick insects and leaf insects	PHASMIDA
stiletto flies	THEREVIDAE (in particular *Thereva* spp.)
stiletto fly, common	*Thereva nobilitata* (Fabricius)
stilt-legged fly	*Micropeza corrigiolata* (L.) syn. *Tylos corrigiolatus* (L.)
St. Mark's flies	BIBIONIDAE
St. Mark's fly	*Bibio marci* (L.)
stone mite	*Petrobia latens* (Müller)
stored nut moth	*Paralipsa gularis* (Zeller) syn. *Aphomia gularis* (Zeller)
stored product psocid	*Liposcelis bostrychophilus* Badonnel ? syn. *L. divinatorius* (Müller) (in part)

COMMON NAMES—Arthropods

Common Name	Scientific Name
stored tobacco moth, *see* warehouse moth	
stout dart moth	*Spaelotis ravida* (Denis & Schiffermüller) syn. *Agrotis ravida* (Denis & Schiffermüller) *Spaelotis obscura* (Brahm)
stouts, *see* clegs, gad flies, horse flies	
strawberry aphid	*Chaetosiphon fragaefolii* (Cockerell) syn. *Capitophorus fragariae* (Theobald) *Pentatrichopus fragaefolii* (Cockerell) *P. fragariae* (Theobald)
strawberry blossom weevil	*Anthonomus rubi* (Herbst)
strawberry fruit weevils, *see* broad-nosed weevil, hairy *and* broad-nosed weevil, smooth	
strawberry ground beetles (*see also* rain beetle)	*Pterostichus madidus* (Fabricius) syn. *Feronia madida* (Fabricius) *Pterostichus melanarius* (Illiger) syn. *Feronia melanarius* (Illiger) misident. *Pterostichus vulgaris* (L.)
strawberry leafhopper	*Aphrodes bicinctus* (Schrank)
strawberry mite	*Phytonemus pallidus* ssp. *fragariae* (Zimmerman)
strawberry rhynchites (weevil)	*Rhynchites germanicus* Herbst syn. *Caenorhinus germanicus* (Herbst)
strawberry root weevils, *see* strawberry weevil, strawberry weevil, scaly *and* strawberry weevils, lesser	
strawberry seed beetle	*Harpalus rufipes* (Degeer) syn. *Ophonus pubescens* (Müller)
strawberry tortrix (moth)	*Acleris comariana* (Lienig & Zeller) syn. *Acalla comariana* (Zeller) *Argyrotoza comariana* (Zeller)
strawberry tortrix (moth), dark	*Olethreutes lacunana* (Denis & Schiffermüller)
†strawberry tortrix parasite (wasp)	*Litomastix aretas* (Walker) syn. *Copidosoma tortricis* Waterstone
strawberry weevil	*Otiorhynchus ovatus* (L.)
strawberry weevil, scaly	*Sciaphilus asperatus* (Bonsdorff) syn. *S. muricatus* (Fabricius)
strawberry weevils, lesser	*Otiorhynchus rugifrons* (Gyllenhal) *Otiorhynchus rugosostriatus* (Goeze)
strawberry whitefly, *see* honeysuckle whitefly	
straw-coloured apple moth, *see* apple moth, straw-coloured	
straw-coloured tortrix (moth)	*Clepsis spectrana* (Treitschke) misindent. *Clepsis constana* (Denis & Schiffermüller)
straw itch mite	*Pyemotes tritici* (La Grèze-Fossat & Montagné) syn. *Pediculoides ventricosus* (Newport) *Pyemotes ventricosus* (Newport)
straw mite, *see* grainstack mite	
striate thrips	*Anaphothrips obscurus* (Müller) syn. *A. striatus* (Osborne)
striped flea beetle, large	*Phyllotreta nemorum* (L.)
striped flea beetle, small	*Phyllotreta undulata* Kutschera
subcutaneous mite, *see* fowl cyst mite	
suckers, *see* psyllids	
sucking lice	ANOPLURA syn. SIPHUNCULATA
sucking lice, biting lice *and* chewing lice	PHTHIRAPTERA
sugar beet thrips, *see* banded glasshouse thrips	

COMMON NAMES—Arthropods

Common Name   **Scientific Name**

*sugar cane borer   (moth) larva of *Opogona sacchari* (Bojer)
summer chafer (beetle)   *Amphimallon solstitialis* (L.)
summer fruit tortrix (moth)   *Adoxophyes orana* (Fischer von Röslerstamm)
     syn. *Capua reticulana* (Hübner)
surface caterpillars, cutworms   (moth) larvae of ⎰ *Agrotis* spp. / *Euxoa* spp. / *Noctua pronuba* (L.)
Surinam cockroach   *Pycnoscelus surinamensis* (L.)
   swallow bug, *see* martin bug
   swarming flies, *see* cluster flies *and* sweat flies
swarming fly, yellow   *Thaumatomyia notata* (Meigen)
     syn. *Chloropisca circumdata* (Meigen)
sweat flies   *Hydrotaea* spp.
   (*see also* cattle sweat fly)
sweat flies, common   ⎰ *Hydrotaea albipuncta* (Zetterstedt) / *Hydrotaea meteorica* (L.) / *Hydrotaea occulta* (Meigen)
swede midge   *Contarinia nasturtii* (Kieffer)
   sweet potato whitefly, *see* tobacco whitefly
swift and swallow parasitic fly   *Crataerina pallida* (Latreille)
sword-grass moth   *Xylena exsoleta* (L.)
sword-grass moth, dark   *Agrotis ipsilon* (Hufnagel)
   larva = black cutworm   misspelling *A. ypsilon* (Hufnagel)
sword-grass moth, red   *Xylena vetusta* (Hübner)
sycamore aphid   *Drepanosiphum platanoidis* (Schrank)
sycamore gall mite   *Aculops acericola* (Nalepa)
sycamore sawfly   *Pristiphora subbifida* (Thomson)
   symbiotic mange mite, *see* chorioptic mange mite
symphylids   SYMPHYLA
*Syrian conifer longhorn beetle   *Ergates faber* (L.)

tanbark borer   (beetle) larva of *Phymatodes testaceus* (L.)
     syn. *P. variabilis* (L.)
     Callidium variabilis (L.)
   tanner beetle, *see* oak longhorn beetles (in part)
*Tanzanian coffee longhorn beetle   *Pachydissus hector* Kolbe
   tapestry moth, *see* clothes moth, white-tip
   tare seed weevil, *see* vetch seed weevil
tarnished plant bug   *Lygus rugulipennis* Poppius
     misident. *L. pratensis* (L.)
tarsonemid mites   TARSONEMIDAE
tawny-barred angle moth   *Semiothisa liturata* (Clerck)
tawny burrowing bee   *Andrena fulva* (Müller in Allioni)
     syn. *A. armata* (Gmelin in Linnaeus)
teasel fly   *Phytomyza ramosa* Hendel
tenebrionid beetles   TENEBRIONIDAE
†ten-spot ladybird (beetle)   *Adalia decempunctata* (L.)
     syn. *A. 10-punctata* (L.)
   terminal rosette-gall midge *see* willow rosette-gall midge
thistle aphid   *Brachycaudus cardui* (L.)
     syn. *Anuraphis cardui* (L.)
thistle root gall beetle   *Cleonus pigra* (Scopoli)
     misspelling *C. piger*
thrips   THYSANOPTERA
throat bot fly   *Gasterophilus nasalis* (L.)
   thunderflies, *see* grain thrips *and* thrips

COMMON NAMES — Arthropods

Common Name	Scientific Name
thuya aphid	*Cinara tujafilina* (del Guercio)
thyme leafhopper	*Emelyanoviana mollicula* (Boheman)
	syn. *Dikraneura mollicula* (Boheman)
thyme moth	*Scrobipalpa artemisiella* (Treitschke)
	syn. *Phthorimaea artemisiella* (Treitschke)
ticks and mites	{ ACARI
	ACARINA
ticks, spiders, mites, harvestmen, etc.	ARACHNIDA
tiger beetles	CICINDELINAE
tiger beetles and ground beetles	CARABIDAE
tiger crane flies, *see* spotted crane flies	
timber borer (weevil)	*Caulotrupodes aeneopiceus* (Boheman)
	syn. *Caulotrupis aeneopiceus* (Boheman)
timberman, common (beetle)	*Acanthocinus aedilis* (L.)
timbermen, *see* longhorn beetles	
timberworm, large	(beetle) larva of *Hylecoetus dermestoides* (L.)
timberworms	(beetle) larvae of LYMEXYLIDAE
timberworm, ship (beetle *larva*), *see* ship timberworm	
timothy flies	*Nanna* spp.
	syn. *Amaurosoma* spp.
timothy tortrix (moth)	*Aphelia paleana* (Hübner)
	syn. *Tortrix paleana* (Hübner)
tin-tack gall mite, *see* nail gall mite	
*tobacco whitefly	*Bemisia tabaci* (Gennadius)
tomato erineum mite	*Aceria lycopersici* (Wolffenstein)
tomato leaf miner	(fly) larva of *Liriomyza bryoniae* (Kaltenbach)
	syn. *L. solani* Hering
tomato moth	*Lacanobia oleracea* (L.)
	syn. *Diataraxia oleracea* (L.)
	Hadena oleracea (L.)
	Mamestra oleracea (L.)
tomato russet mite	*Aculops lycopersici* (Massee)
	syn. *Aculus lycopersici* (Massee)
	Phyllocoptes lycopersici Massee
tomato-worm, *see* Old World bollworm	
†tombstone chalcids	(wasp) pupae of { *Eulophus pennicornis* Nees
	Eulophus larvarum (L.)
tortoise beetles	*Cassida* spp.
*tortoise tick	*Hyalomma aegyptium* (L.)
tortrices (moths)	TORTRICIDAE
larvae = leaf rollers	
tortrix moths, *see* tortrices	
††tree damsel bug	*Himacerus apterus* (Fabricius)
	syn. *Nabis apterus* (Fabricius)
tree-hole mosquito	*Anopheles plumbeus* Stephens
tree wasp	*Dolichovespula sylvestris* (Scopoli)
	syn. *Vespula sylvestris* (Scopoli)
trefoil flower midge	*Contarinia loti* (Degeer)
trefoil leaf weevil	*Hypera punctata* (Fabricius)
	syn. *Phytonomus punctata* (Fabricius)
	? syn. *Hypera austriaca* (Schrank)
	H. zoilus (Scopoli)
	Phytonomus austriacus (Schrank)
tropical brown dog tick, *see* kennel tick	
tropical rat mite	*Ornithonyssus bacoti* (Hirst)
	syn. *Bdellonyssus bacoti* (Hirst)
	Liponyssus bacoti (Hirst)

COMMON NAMES — Arthropods

Common Name   **Scientific Name**

tropical warehouse moth .. .. .. *Ephestia cautella* (Walker)
    syn. *E. defectella* (Walker)
true lovers knot moth .. .. .. *Lycophotia porphyrea* (Denis & Schiffermüller)
    syn. *Peridroma porphyrea* (Denis & Schiffermüller)
  tubercle-bearing louse, *see* blue cattle louse
  tufted apple-bud moth, *see* apple-bud moth, tufted
tulip bulb aphid .. .. *Dysaphis tulipae* (Boyer de Fonscolombe)
    syn. *Anuraphis tulipae* (Boyer de Fonscolombe)
    Dentatus tulipae (Boyer de Fonscolombe)
    Sappaphis tulipae (Boyer de Fonscolombe)
  turf ant, *see* yellow meadow ant
turkey louse, large .. .. *Chelopistes meleagridis* (L.)
    syn. *C. stylifer* (Nitzsch)
turkey wing louse .. .. *Oxylipeurus polytrapezius* (Burmeister)
Turkish grain beetle .. .. *Cryptolestes turcicus* (Grouvelle)
turnip flea beetles .. .. *Phyllotreta atra* (Fabricius)
    Phyllotreta consobrina (Curtis)
    Phyllotreta cruciferae (Goeze)
    Phyllotreta nigripes (Fabricius)
  turnip 'fly', *see* turnip flea beetles
turnip gall weevil .. .. *Ceutorhynchus pleurostigma* (Marsham)
turnip moth .. .. *Agrotis segetum* (Denis & Schiffermüller)
  larva = common cutworm
turnip mud beetles .. .. *Helophorus porculus* Bedel
    syn. *Megempleurus porculus* (Bedel)
    misident. *Helophorus rugosus* Olivier
    Helophorus rufipes (Bosc d'Antic)
    syn. *H. rugosus* Olivier
    Megempleurus rugosus (Olivier)
turnip root fly .. .. *Delia floralis* (Fallén)
    syn. *Erioischia floralis* (Fallén)
    Phorbia floralis (Fallén)
turnip sawfly .. .. *Athalia rosae* (L.)
    syn. *A. colibri* (Christ)
turnip stem weevil .. .. *Ceutorhynchus contractus* (Marsham)
tussock moths .. .. LYMANTRIIDAE
twentyfour-spot ladybird (beetle) .. .. *Subcoccinella vigintiquattuorpunctata* (L.)
    syn. *S. 24-punctata* (L.)
  twenty plumed moth, *see* many plumed moth
†twentytwo-spot ladybird (beetle) .. .. *Thea vigintiduopunctata* (L.)
    syn. *Psyllobora vigintiduopunctata* (L.)
    Thea 22-punctata (L.)
  twig cutting weevils, *see* rhynchites
two-banded fungus beetle .. .. .. *Alphitophagus bifasciatus* (Say)
  two-spot carpet beetle, *see* fur beetle
†two-spot ladybird (beetle) .. .. .. *Adalia bipunctata* (L.)
    syn. *A. 2-punctata* (L.)
two-spotted spider mite .. .. .. *Tetranychus urticae* Koch
    syn. *T. bimaculatus* Harvey
    ? misident. *T. telarius* (L.)
†two-spotted spider mite predator (mite) .. *Phytoseiulus persimilis* Athias-Henriot
    syn. *P. riegeli* Dosse
  two-toothed pine beetle, *see* pine beetle, two-toothed

COMMON NAMES—Arthropods

Common Name	Scientific Name
upland click beetles, *see* click beetles, upland	
upland wireworms	(beetle) larvae of *Ctenicera* spp. syn. *Corymbites* spp.
urinal fly	*Teichomyza fusca* Macquart
vapourer moth	*Orgyia antiqua* (L.)
varied carpet beetle	*Anthrenus verbasci* (L.)
variegated cutworm	(moth) larva of *Peridroma saucia* (Hübner) syn. *Agrotis saucia* (Hübner) *Peridroma margaritosa* (Haworth) misident. *P. porphyrea* (Denis & Schiffermüller)
variegated golden tortrix (moth)	*Archips xylosteana* (L.)
velvet ants	MUTILLIDAE
vetch aphid	*Megoura viciae* Buckton
vetch leaf midge	*Dasineura viciae* (Kieffer)
vetch seed weevil	*Apion pomonae* (Fabricius)
viburnum aphid	*Aphis viburni* Scopoli
viburnum beetle	*Pyrrhalta viburni* (Paykull) syn. *Galerucella viburni* (Paykull)
viburnum tortrix (moth)	*Acleris schalleriana* (L.)
viburnum whitefly	*Aleurotuba jelinekii* (von Frauenfeld) syn. *Aleurotrachelus jelinekii* (von Frauenfeld)
vine erineum mite, *see* vine leaf blister mite	
vinegar barrel psocid, *see* pale booklouse	
vinegar flies, *see* fruit flies, small	
vine leaf blister mite	*Colomerus vitis* (Pagenstecher) syn. *Eriophyes vitis* (Pagenstecher) *Phytoptus vitis* Pagenstecher
vine louse, *see* grape phylloxera	
*vine mealybug	*Pseudococcus maritimus* (Ehrhorn)
(*see also* glasshouse mealybug)	
vine moth	*Eupoecilia ambiguella* (Hübner)
vine tortrix (moth), *see* fruit tree tortrices	
vine weevil	*Otiorhynchus sulcatus* (Fabricius) syn. *Brachyrhinus sulcatus* (Fabricius)
viola sawfly	*Protemphytus pallipes* (Spinola) syn. *Ametastegia pallipes* (Spinola) *Emphytus pallipes* (Spinola) misspelling *Protoemphytus pallipes* (Spinola)
violet aphid	*Myzus ornatus* Laing
†violet ground beetle	*Carabus violaceus* L.
violet leaf midge	*Dasineura affinis* (Kieffer)
violet tanbark beetle	*Callidium violaceum* (L.)
violet willow beetle	*Helops caeruleus* (L.)
vole and shrew tick	*Ixodes trianguliceps* Birula syn. *Ixodes tenuirostris* Neumann
Walker's heath tortrix (moth)	*Philedonides lunana* (Thunberg) syn. *Philedone prodromana* (Hübner)
wall-louse, *see* bed bug	
walnut aphid, large	*Callaphis juglandis* (Goeze)
walnut aphid, small	*Chromaphis juglandicola* (Kaltenbach)

COMMON NAMES—Arthropods

Common Name	Scientific Name
walnut leaf gall mite	*Aceria tristriatus* (Nalepa)
	syn. *Eriophyes tristriata* (Nalepa)
waltzing midges, *see* moth flies	
warble flies, *see* ox warble flies	
warehouse moth	*Ephestia elutella* (Hübner)
	syn. *E. sericarium* (Scott)
wasp beetle	*Clytus arietis* (L.)
wasp, common	*Vespula vulgaris* (L.)
	syn. *Paravespula vulgaris* (L.)
	Vespa vulgaris L.
wasps	VESPIDAE
wasps, ants, bees, sawflies etc.	HYMENOPTERA
waste grain beetle, *see* two-banded fungus beetle	
watercress beetle, *see* mustard beetles	
water-lily aphid	*Rhopalosiphum nymphaeae* (L.)
water-lily beetle	*Galerucella nymphaeae* (L.)
wax moth, bumble-bee	*Aphomia sociella* (L.)
wax moth, *see* honeycomb moth	
wax moth, lesser	*Achroia grisella* (Fabricius)
	syn. *Meliphora grisella* (Fabricius)
webbing clothes maggot	(moth) larva of *Tineola bisselliella* (Hummel)
web-spinning larch sawfly	*Cephalcia lariciphila* (Wachtl)
	misident. *C. alpina* (Klug)
weevils	CURCULIONIDAE
Welsh chafer (beetle)	*Hoplia philanthus* (Fuessly)
*West African ghoon beetle	*Dinoderus bifoveolatus* (Wollaston)
*West African wood borer	(beetle) larva of *Xyloperthodes nitidipennis* (Murray)
*western flower thrips	*Frankliniella occidentalis* (Pergande)
*western pine beetle	*Dendroctonus brevicomis* LeConte
western stout pillbug (woodlouse)	*Armadillidium depressum* Brandt
wet grain mite	*Caloglyphus berlesei* (Michael)
Weymouth pine adelges	*Pineus strobi* (Hartig)
	syn. *P. strobus* (Ratzeburg)
wharf borer	(beetle) larva of *Nacerdes melanura* (L.)
	syn. *Anoncodes melanura* (L.)
wheat blossom midge, orange	*Sitodiplosis mosellana* (Géhin)
†wheat blossom midge parasite (wasp)	*Leptacis tipulae* (Kirby)
wheat blossom midges, *see* wheat blossom midge, orange	
and wheat blossom midge, yellow	
wheat blossom midge, yellow	*Contarinia tritici* (Kirby)
wheat bulb fly	*Delia coarctata* (Fallén)
	syn. *Leptohylemyia coarctata* (Fallén)
	syn. and misspelling *Hylemyia coarctata* (Fallén)
wheat cutworm	(moth) larva of *Apamea sordens* (Hufnagel)
wheat flea beetle	*Crepidodera ferruginea* (Scopoli)
wheat mud beetle, *see* wheat shoot beetle	
wheat-pollard itch mite, *see* scaly grain mite	
wheat shoot beetle	*Helophorus nubilus* Fabricius
	syn. *Empleurus nubilis* (Fabricius)
wheat stem sawfly	*Cephus pygmeus* (L.)
	misspelling *Cephus pygamaeus*
†whirligig mites	*Anystis* spp.
white blind springtails	*Onychiurus* spp.
white butterfly, green-veined	*Pieris napi* (L.)
white butterfly, large	*Pieris brassicae* (L.)
white butterfly, small	*Pieris rapae* (L.)
	syn. *Artogeia rapae* (L.)

COMMON NAMES—Arthropods

Common Name	Scientific Name
†white butterfly parasitic wasp	*Pteromalus puparum* (L.)
white clover seed midge, *see* clover seed midge, white	
white clover seed weevil, *see* clover seed weevil, white	
white ermine moth	*Spilosoma lubricipede* (L.)
whiteflies	ALEYRODIDAE syn. ALEURODIDAE
whiteflies, aphids, hoppers, mealybugs, phylloxeras, psyllids, scale insects	HOMOPTERA
white grubs	(beetle) larvae of { *Melolontha hippocastani* Fabricius / *Melolontha melolontha* (L.) syn. *M. vulgaris* Fabricius }
white-legged black millepede	*Tachypodoiulus niger* (Leach) syn. *T. albipes* (Koch)
white-line dart moth	*Euxoa tritici* (L.)
white-marked spider beetle	*Ptinus fur* (L.)
white peach scale, *see* peach scale, white	
white-shouldered house moth	*Endrosis sarcitrella* (L.) syn. *E. lactella* (Denis & Schiffermüller)
white-tip clothes moth, *see* clothes moth, white-tip	
white wave moth, common	*Cabera pusaria* (L.)
whooping gall midge, *see* juniper gall midge	
wicker longhorn beetles	{ *Gracilia minuta* (Fabricius) / **Nathrius brevipennis* (Mulsant) syn. *Leptideella brevipennis* (Mulsant) }
wild service aphid	*Dysaphis aucupariae* (Buckton) syn. *Sappaphis aucupariae* (Buckton)
willow aphid, black	*Pterocomma salicis* (L.) syn. *Melanoxantherium salicis* (L.)
willow aphid, large	*Tuberolachnus salignus* (Gmelin) syn. *Pterochlorus salicis* (Sulzer) *P. saligna* (Gmelin)
willow aphids	*Cavariella* spp.
willow aphid, small	*Aphis farinosa* Gmelin syn. *A. saliceti* Kaltenbach
willow bark phylloxera	*Phylloxerina salicis* (Lichtenstein) syn. *Phylloxera salicis* (Lichtenstein)
willow bean-gall sawfly	*Pontania proxima* (Lepeletier) syn. *Nematus proximus* Lepeletier
†willow bean-gall sawfly predator (weevil)	*Curculio salicivorus* Paykull syn. *Balanobius salicivorus* (Paykull)
willow beetle, *see* osier weevil	
willow borer	(beetle) larva of *Cryptorhynchus lapathi* (L.)
willow boring sawfly, *see* willow wood wasp	
willow bud sawfly	*Euura mucronata* (Hartig) misident. *E. saliceti* (Fallén)
willow button-top midge	*Rhabdophaga heterobia* (Löw)
willow—carrot aphid	*Cavariella aegopodii* (Scopoli)
willow ermine moth	*Yponomeuta rorrella* (Hübner)
willow flea beetle	*Chalcoides aurata* (Marsham) syn. *Crepidodera aurata* (Marsham) misident. *C. helxines* (L.)
willow froghopper	*Aphrophora salicina* (Goeze) syn. *A. salicis* (Degeer)
willow leaf-folding midge	*Rhabdophaga clausilia* (Bremi) syn. *Dasineura clausilia* (Bremi)
willow leaf gall sawflies	*Pontania* spp.

COMMON NAMES—Arthropods

Common Name	Scientific Name
willow—parsnip aphid	*Cavariella theobaldi* (Gillette & Bragg)
willow pea-gall sawflies	*Pontania pedunculi* (Hartig) *Pontania viminalis* (L.) syn. *Nematus viminalis* (L.) *Pontania harrisoni* Benson
willow rosette-gall midge	*Rhabdophaga rosaria* (Loew)
willow sawfly	*Nematus salicis* (L.) syn. *Pteronidea salicis* (L.)
willow sawfly, lesser	*Nematus pavidus* Lepeletier syn. *Pteronidea pavida* (Lepeletier)

willow scale	*Chionaspis salicis* (L.)
	  syn. *C. alni* Signoret
	    C. populi (Baerensprung)

willow scurfy scale, *see* willow scale	
willow shoot moth	*Agonopterix conterminella* (Zeller) syn. *Depressaria conterminella* Zeller
willow shot-hole midges	*Helicomyia saliciperda* (Dufour) (also known as willow wood midge) *Rhabdophaga justini* Barnes *Rhabdophaga purpureaperda* Barnes *Rhabdophaga triandraperda* Barnes
willow stem gall midge	*Rhabdophaga salicis* (Schrank)
willow stem sawfly	*Janus luteipes* (Lepeletier)
willow terminal leaf midge	*Rhabdophaga terminalis* (Loew)

willow tortrix (moth)	*Epinotia cruciana* (L.)
	  syn. *Eucosma cruciana* (L.)
	    Panoplia cruciana (L.)

willow wood midge, *see* willow shot-hole midge (in part)	
willow wood wasp	*Xiphydria prolongata* (Fourcroy)
window flies	SCENOPINIDAE (in particular *Scenopinus* spp.)
window gnat, common	*Sylvicola fenestralis* (Scopoli)

window gnats, sewage filter-bed flies	ANISOPODIDAE
	  (in particular *Sylvicola* spp.
	    syn. *Anisopus* spp.)

wingless weevils	*Otiorhynchus* spp. misspelling *Otiorrhynchus*
wing wagger flies	SEPSIDAE (in particular *Sepsis* spp.)
winter gnat	*Trichocera saltator* (Harris)
winter grain mite, *see* red-legged earth mite	
winter moth	*Operophtera brumata* (L.) syn. *Cheimatobia brumata* (L.)
winter moth, northern	*Operophtera fagata* (Scharfenberg)
winter moth, small, *see* winter moth	
wire centipedes	GEOPHILIDAE

wireworms	(beetle) larvae of ELATERIDAE
	  in particular *Agriotes lineatus* (L.)
	    Agriotes obscurus (L.)
	    Agriotes sputator (L.)
	    Athous spp.

witches' broom mite (of birch)	*Acalitus rudis* (Canestrini)
	  syn. *Aceria rudis* (Canestrini)
	    Eriophyes rudis (Canestrini)
	    Phytoptus rudis Canestrini

witches' broom mite (of willow)	*Eriophyes triradiatus* (Nalepa)
	  syn. *Aceria triradiatus* (Nalepa)
	    Phytoptus triradiatus Nalepa

wolf moth, *see* corn moth	
wood ant	*Formica rufa* L.

COMMON NAMES—Arthropods

Common Name	Scientific Name
wood ant, hairy	*Formica lugubris* Zetterstedt
wood borers	(beetle) larvae of BOSTRICHIDAE
wood boring weevils, *see* Chilean wood-boring weevil *and* New Zealand wood-boring weevils	
woodland floor millepede	*Cylindroiulus punctatus* (Leach)
wood leopard moth, *see* leopard moth	
woodlice	ISOPODA
†wood wasp parasite (wasp)	*Rhyssa persuasoria* (L.)
woodworm	(beetle) larva of *Anobium punctatum* (Degeer)
	syn. *A. domesticum* (Fourcroy)
	misident. *A. striatum* Fabricius
woodworms	(beetle) larvae of ANOBIIDAE
woolly aphid	*Eriosoma lanigerum* (Hausmann)
†woolly aphid parasite (wasp)	*Aphelinus mali* (Haldeman)
woolly aphis parasite (wasp), *see* woolly aphid parasite	
woolly bears	(beetle) larvae of *Anthrenus* spp.
woolly currant scale	*Pulvinaria ribesiae* Signoret
woolly poplar-stem aphid	*Phloeomyzus passerinii* (Signoret)
woolly vine scale	*Pulvinaria vitis* (L.)
Worthing phorid, *see* mushroom scuttle flies	
†wrinkled rove beetle	*Anotylus rugosus* (Fabricius)
†xantholine rove beetles	*Xantholinus* spp.
yellow ant	*Lasius mixtus* (Nylander)
yellow bamboo longhorn beetle, *see* bamboo longhorn beetle, yellow	
*yellow-bowed longhorn beetle	*Plagionotus arcuatus* (L.)
yellow cereal fly, *see* cereal fly, yellow	
yellow dung fly, *see* dung fly, yellow	
yellow flower thrips	*Thrips flavus* Schrank
	syn. *T. frankeniae* Bagnall
yellow meadow ant	*Lasius flavus* (Fabricius)
yellow mealworm beetle, *see* mealworm beetle, yellow	
†yellow ophion	*Ophion luteus* (L.)
yellow orchid aphid, *see* orchid aphid, yellow	
yellow orchid thrips, *see* orchid thrips, yellow	
yellow swarming fly, *see* swarming fly, yellow	
yellow-tail moth	*Euproctis similis* (Fuessly)
	misident. *E. chrysorrhoea* (L.)
yellow thrips, *see* orchid thrips, yellow	
yellow underwing moth, large	*Noctua pronuba* (L.)
larva = cutworm	syn. *Triphaena pronuba* (L.)
surface caterpillar	
yellow underwing moth, lesser	*Noctua comes* (Hübner)
	syn. *Euschesis comes* (Hübner)
yellow v moth	*Oinophila v-flava* (Haworth)
yellow wheat blossom midge, *see* wheat blossom midge, yellow	
yew big bud mite, *see* yew gall mite	
yew gall midge	*Taxomyia taxi* (Inchbald)
yew gall mite	*Cecidophyopsis psilaspis* (Nalepa)
	syn. *Eriophyes psilaspis* (Nalepa)
	Phytoptus psilaspis Nalepa

COMMON NAMES—Arthropods

Common Name   **Scientific Name**

yew scale   *Parthenolecanium pomeranicum* (Kawecki)
                                        syn. *Eulecanium pomeranicum* (Kawecki)
                                        misident. *Lecanium cornicrudum* (Green)

Zeller's midget moth   *Phyllonorycter messaniella* (Zeller)

Common—Scientific Names of Molluscs

Common Name	Scientific Name
apple, snail, *see* Roman snail	
black slug	*Arion ater* (L.)
banded snail, larger	*Cepaea nemoralis* (L.)
	syn. *Helix nemoralis* L.
banded snail, smaller	*Cepaea hortensis* (Müller)
	syn. *Helix hortensis* (Müller)
cellar slug	*Limax flavus* L.
chestnut slug	*Deroceras panormitanum* (Lessona & Pollonera)
	syn. *D. caruanae* (Pollonera)
common shipworm, *see* shipworm, common	
common snail, *see* garden snail	
dairy slug, *see* cellar slug	
dusky slug	*Arion subfuscus* (Draparnaud)
dwarf pond snail	*Lymnaea truncatula* (Müller)
	syn. *Limnaea truncatula* (Müller)
field slug	*Deroceras reticulatum* (Müller)
	syn. *Agriolimax reticulatus* (Müller)
false shipworm	*Xylophaga dorsalis* Turton
garden slugs	{ *Arion distinctus* (Mabille)
	Arion hortensis Férussac (also known as yellow-soled slug)
garden snail	*Helix aspersa* Müller
glasshouse slug	*Limax valentianus* Férussac
great slug	*Limax maximus* L.
grey field slug, *see* field slug	
heath snail	*Helicella itala* (L.)
hedgehog slug	*Arion intermedius* Normand

COMMON NAMES—Molluscs

Common Name	Scientific Name

keeled slugs	{ *Milax gagates* (Draparnaud)
	Tandonia budapestensis (Hazay)
	  syn. *Milax budapestensis* (Hazay)
	  M. gracilis (Leydig)
	Tandonia sowerbyi (Férussac)
	  syn. *Milax sowerbyi* (Férussac)

larger banded snail, *see* banded snail, larger

marsh slug	*Deroceras laeve* (Müller)
mud snail	*Lymnaea glabra* (Müller)

(*see also* dwarf pond snail)

netted slug, *see* field slug

piddocks	PHOLADIDAE

pointed helicellid, *see* pointed snail

pointed snail	*Cochlicella acuta* (Müller)
red slug	*Arion rufus* (L.)
Roman snail	*Helix pomatia* L.
shipworms, slugs, snails, etc.	MOLLUSCA
shipworms	TEREDINIDAE (in particular *Teredo* spp.)
shipworm, common	*Nototeredo norvagicus* (Spengler) syn. *Teredo norvagicus* (Spengler)
silver slug	*Arion silvaticus* Lohmander
slugs, snails, shipworms, etc.	MOLLUSCA

smaller banded snail, *see* banded snail, smaller

snails, shipworms, slugs, etc.	MOLLUSCA
strawberry snail	*Trichia striolata* (Pfeiffer) syn. *Hygromia striolata* (Pfeiffer)
striped snail	*Cernuella virgata* (da Costa) syn. *Helicella virgata* (da Costa)

subterranean slugs, *see* keeled slugs

white-soled slugs	{ *Arion circumscriptus* Johnston *Arion fasciatus* (Nilsson)

(*see also* silver slug)

†worm-eating slugs	*Testacella* spp.
worm slug	*Boettgerilla pallens* Simroth

yellow slug, *see* cellar slug
yellow-soled slug, *see* garden slugs (in part)